CONSERVATION EC

Wye College, University of London

4-5 September 1992

Hadlow
college

H

Date of Return	Date of Return	Date of Return
1 9 APR 2010	1 4 JAN 2013	
ONE WEEK LOAN 2 1 APR 2010	2 1 NOV 2015	
- 5 MAY 2010	1 6 NOV 2016	
- 8 MAR 2011		
ONE WEEK LOAN 1 5 MAR 2011		
2 2 MAR 2011		
2 2 JUN 2011		
2 6 MAR 2012		
- 8 NOV 2012		

2

First published in the United Kingdom by
Wye College Press
Wye College, University of London,
Wye, Ashford, Kent TN25 5AH

© Wye College Press, 1994

British Library Cataloguing-in-Publication Data
A catalogue record for this book is available from the British Library

ISBN 0-86266-037-8

Cover photograph by N. R .Bannister

Typeset by
University of London Computer Centre
Phototypesetter Service

Printed in Great Britain by
Headley Brothers Ltd.,
Ashford, Kent

Contents

PREFACE

Hedgerows, the most frequently occurring semi-natural features on lowland farms in Britain, increasingly represent the few remaining refuges where wildlife can survive an otherwise intensively managed landscape. In the 1960s and 1970s, long before the days of agricultural set-aside and/or the emergence of a widely shared ethic of countryside conservation, the importance of hedges as wildlife habitats and corridors was recognized in research carried out at Monks Wood Experimental Station and in the publication of the classic New Naturalist volume *Hedges* in 1974. This same volume drew attention to the losses of field boundaries which were then taking place as field sizes were expanded to increase agricultural efficiency.

Today it is perhaps something of a surprise to learn that, despite the background of increasing food surpluses in Europe and the limiting returns implied by still greater agricultural efficiency, hedgerow losses have actually accelerated during the past decade. The flux of gains and losses is more revealing, showing that although, increasingly, new hedges are being planted, greater lengths of existing hedges are being removed. More worrying is the state of management of older hedges, which have become gappy and fragmented through neglect, or have reverted to lines of tall shrubs or trees.

The continuing vulnerability of hedgerows in the 1990s, and the need to update and consolidate hedgerow research carried out since the mid-1970s, provided the focus for a meeting of the Conservation Ecology group of the British Ecological Society. This was held at Wye College, University of London during 4-5 September 1992, attracting a wide range of contributors whose papers are summarized here. The proceedings were enlivened by an afternoon visit to Great Tong Farm near Headcorn, Kent, one of three Link Farms set up by the Kent Farming and Wildlife Advisory Group to demonstrate the interface between conservation and sound farming practice. We are grateful to Mrs Roz Day and to Mr Paul Cobb, the Kent FWAG adviser, for making this visit possible.

Finally, we would like to thank the British Ecological Society for their support of this meeting, and our many helpers on the day, including Nikki Bannister, whose idea it was to hold the meeting, Mary Berry, Frances Burch, Anne-Maria Brennan, Jeff Brooks, and Andrew Pennington and also Andrea Odon, Pam Rogers and Mike Wilkinson from Wye College Press.

Trudy Watt
Peter Buckley

Wye, July 1994

CHAPTER 1

INTRODUCTION

M.D. Hooper
Swales, Pilton, Peterborough, Cambridgeshire PE8 5SN

Hedges have been of interest for a long time. The earliest papers, specifically upon hedges and which can be called scientific, all date from the period when many hedges were being planted in England. Perhaps not surprisingly, all were written in an agricultural context and all advanced the view that hedges were too much of a good thing: there were too many hedges and many hedges were too big (Stiva, 1800; Cambridge, 1845; Grant, 1845; Grigor, 1845; Turner, 1845).

As if to redress the balance and to emphasize that hedges had benefits a second period of scientific interest, still in an agricultural context and roughly up to 1960, concentrated on shelter effects, on both stock and crops (Jensen, 1954; Caborn, 1965). Nevertheless, by about 1960 it was being realized that shelter was not universally of benefit to agriculture (Lewis, 1965; Marshall, 1967) and interest began to move to the conservation values of hedges.

In 1960 the then Nature Conservancy set up the Toxic Chemicals and Wildlife Section of its Research Branch under Dr Norman Moore. The remit of this section was originally to assess the effects of agricultural chemicals upon wildlife (later it included all pollutant chemicals). To carry out this remit effectively Dr Moore realized that his section would also have to have some assessment of the impact of the other changes in the rural environment. Among these other changes, such as the increased drainage of land and the ploughing up of pasture, he included the destruction of hedgerows. Dr Moore and Dr Brian Davis therefore began to assess the rate and distribution of loss of hedges and, to get some idea of the consequent impact upon wildlife, began to count the nesting bird population in hedges.

By the time I joined the section in 1963 they had reasonable evidence of the distribution of hedgerow loss and reasonable assessments of the bird diversity in hedges under different management regimes (Moore *et al.*, 1967), but they had also found that the woody plant diversity of the hedges affected the bird populations. The more kinds of trees and shrubs a unit length of hedge had the more likely it was to be rich in bird species. Hence for the Nature Conservancy, wishing to advise farmers on how to manage hedges (assuming at least some might be left), the question was what causes one hedge to be pure hawthorn and another to have a mixture of trees and shrubs on it? The three people in the team looking at hedges, Dr Moore, Dr Davis and Dr Ernie Pollard (who had also joined in 1963) were all zoologists so I, as a botanist, was asked to look at the possible causes of plant diversity in hedges. Was it management or did the answer lie in the soil?

I fairly quickly found that, although management regimes and soil types had effects on the frequency of individual species, overall diversity remained unaffected. Then from reading *The Flora of Wiltshire* (Grose, 1957) I got the idea, not only that the older a hedge was the richer it was in trees and shrubs, but that the increase in number of species was a regular progression of one new species each hundred years. Evidence in support of this idea was found by enlisting the aid of Professor W.G. Hoskins who produced the probable date of origin from documentary evidence of about sixty Devon hedges in which the numbers of tree and shrub species could be determined. Subsequently hedges' ages and floristics were recorded in other counties and the idea of one new species for each century seemed confirmed for southern England except for the borderland with Wales where a tradition of planting mixed hedges obscured any relationship (Hooper, 1970a, 1971).

Over the same period Dr Pollard looked at the diversity of plants and insects in the hedge bottom (Pollard, 1968) and small mammals (Pollard and Relton, 1970) and tested the idea that landscape diversity could affect population stability (Pollard, 1971). He also showed that many ancient hedges which still survived could be rich in species, not because of colonization over time, but because they were relics of former woodland (Pollard, 1973).

These hedgerow studies at Monks Wood virtually came to a standstill (though they were later resumed both at Monks Wood and Merlewood), with the formation of the Institute of Terrestrial Ecology in 1973.

Studies elsewhere at first followed three lines: (a) looking at the possibility of dating hedges in a local history context, (b) at the importance of hedges to birds and (c) studies on the influence of hedges on invertebrate populations at the Game Conservancy at Southampton (Hickman, 1994; MacLeod, 1994; Rothery, 1994). The historical studies met with mixed success, (Cameron and Pannett, 1980; Hewlett, 1973; Hussey, 1987; Roseman, 1991) and while it is now generally accepted by local historians that tree and shrub diversity in hedges is good evidence of age, it is also accepted that one species for each century is not a rule to be followed slavishly.

Similarly, the bird work still remains inconclusive. Birds in hedges have been studied for many years (Alexander, 1932; Chapman, 1939) but it was not until the mid-1960s that serious statistical study began (Moore et al., 1967). Unfortunately, this was when bird populations were beginning to recover from the harsh winter of 1962-63. Some studies from this period have not allowed for this and are therefore suspect. There seems to me to be three main questions:

- Whether the species depend for their continuation through time on hedges?
- Whether bird diversity on farmland is dependent upon the frequency of hedges in the landscape?
- Whether the bird abundance and diversity in an individual hedge is dependent upon the management of that hedge?

The answer to the first question must be that it depends upon the species. It appears that some birds use hedges as a habitat only when a preferred habitat is full. This has been reported for blackbirds (Parslow, 1969), tits (Krebs, 1971), wrens (Williamson, 1969) and wood pigeons (Murton and Westwood, 1974).

Nevertheless, experiments by Edwards (1977) show that, for some species, the farmland habitat is filled to saturation and that there is an excess, floating, population unable to acquire territories. Hence, taking the first view, that hedges depend for their population upon, say, the excess production of young in woods, one could say that the species depend upon the woodland rather than upon hedges. Yet Edwards' work suggests that hedge destruction would reduce the carrying capacity of farmland and, even accepting the hedge habitat as suboptimal, hedge removal could reduce the probability of offspring in the future.

Edwards' work, however, was carried out in an area of few hedges and it is possible that landscape diversity is significant here. Hooper (1970*b*) noted that the number of birds nests per unit length of hedge varied with the number of hedges in the immediate neighbourhood of that hedge, and upon the quality of those hedges. Thus a single hawthorn hedge in an otherwise bare landscape or a single hawthorn hedge surrounded by elm hedges, had a high number of nests in a unit length. The same hedge, surrounded by many hedges of a similar type, would have a relatively low number of nests in a unit length. Hooper, therefore, suggested that above a certain hedge frequency or density (in km of hedge per square km) the factor limiting bird populations could be the food supply, and variation in bird numbers would therefore depend upon cropping patterns and farming practices. Below this critical hedge density the factor limiting bird populations on farmland could be competition for nesting sites in the hedge. Thus hedge removal down to the critical level would have no effect on the bird populations in the hedges (except to force them to nest closer together) but, once the critical level was past, hedge removal would have severe effects on the numbers of birds and bird diversity. It may be that Edwards was working in an area below this threshold while others who deny effects of hedgerow removal (Murton and Westwood, 1974) were working above it. Unfortunately, although Hooper (1970*b*) provides a little evidence, this proposition has never been adequately tested.

The great bulk of recent work (Arnold, 1983; O'Connor and Shrubb, 1986; Osborne, 1982, 1984) has gone to answer the third question on management of individual hedges. Their conclusions are supported and contended by the most recent work carried out at Monks Wood (Parish *et al.*, 1994).

In broad terms, the volume of woody material has greatest influence on the number of species of birds in a hedge, but this is considerably enhanced by the presence, size and management of other components of the field boundary. There appears to be a more or less linear relationship between hedge 'size', especially height, and the number of bird species. The number of woody species, tree numbers and tree height, field size and land use also help improve the explanation of variation in bird numbers. But even the studies on birds which have produced the largest number of papers on hedges still leave questions unanswered.

Indeed it is perhaps only in the field of the statistics of hedgerow losses that definitive, categorical, answers to questions can be given with a known degree of accuracy thanks to studies at ITE Merlewood (Barr *et al.,* 1986, 1991; Barr and Parr, 1994).

After thirty years of studying hedges, I am more conscious of what I do not know and what I am not sure about than the extent of my knowledge. Why are some hedges rich in tree and shrub species and others not? Even if we assume a hedge acquires species as it ages, what is the mechanism? Is it succession, are there changes in the soil which allow species in only after a period of time has elapsed, or is it the rate of death of originally planted individual quickthorns which controls the entry of new shrub species? And what controls which new shrub species? Is that controlled by the frequency of the species in the neighbourhood and the attractiveness of its fruit to birds?

The final result of thirty years is that I can ask more questions than the one I was asked when I started. I hope this symposium will help answer many of them.

References

Alexander WB. (1932) The bird population of an Oxfordshire farm. *Journal of Animal Ecology* **1,** 58-64.

Arnold GR. (1983) The influence of ditch and hedgerow structure, length of hedgerows and area of woodland on bird numbers on farmland. *Journal of Applied Ecology* **20,** 731-50.

Barr C, Benefield C, Bunce R, Ridsdale H, Whittaker M. (1986) *Landscape changes in Britain.* Swindon: Natural Environment Research Council.

Barr C, Howard D, Bunce R, Gillespie M, Hallam C. (1991) *Changes in hedgerows in Britain.* Unpublished Report to Department of Environment, October 1991.

Barr C, Parr TW. (1994) Hedgerows: linking ecological research and countryside policy. In: Watt TA, Buckley GP, eds. *Hedgerow management and nature conservation.* Wye College Press, Wye College, University of London, 119-36.

Caborn M. (1965) *Shelterbelts and windbreaks.* London: Faber and Faber.

Cambridge W. (1845) On the advantages of reducing the size and number of hedges. *Journal of the Royal Agricultural Society* **6,** 333-42.

Cameron RD, Pannett DJ. (1980) Hedgerow shrubs and landscape history in the West Midlands. *Arboricultural Journal* **4,** 147-52.

Chapman WMM. (1939) The bird population of an Oxfordshire farm. *Journal of Animal Ecology* **8,** 286-99.

Edwards PJ. (1977) Re-invasion by farmland bird species following capture and removal. *Polish Ecological Studies* **3,** 53-70.

Grant J. (1845) Remarks on the large hedges of Devonshire. *Journal of the Royal Agricultural Society* **5,** 420-29.

Grigor J. (1845) On fences. *Journal of the Royal Agricultural Society* **6,** 194-228.

Grose D. (1957) *The Flora of Wiltshire.* Devizes.

Hewlett G. (1973) Reconstructing an historical landscape from field and documentary evidence: Otford in Kent. *Agricultural History Review* **21,** 94-110.

Hickman J. (1994) Abstract. Use of *Phacelia tanacetifolia* borders to enhance hoverfly populations in winter wheat. In: Watt TA, Buckley GP, eds. *Hedgerow management and nature conservation.* Wye College Press, Wye College, University of London, 158.

Hooper MD. (1970a) Dating hedges. *Area* **4,** 3-5.

Hooper MD. (1970b) Hedges and birds. *Birds* **3,** 114-17.

Hooper MD. (1971) Hedges and local history. Standing Conference for Local History. London: National Council for Social Services.

Hussey T. (1987) Hedgerow history. *The Local Historian* **17**, 327-42.

Jenson M. (1954) *Shelter effect: the aerodynamics of shelter and its effect on climate and crops.* Copenhagen.

Krebs JR. (1971) Territory and breeding density in the Great Tit, *Parus major. Ecology* **52**, 1-22.

Lewis T. (1965) The effect of shelter on the distribution of insect pests. *Scientific Horticulture* **17**, 74-84.

MacLeod A. (1994) Abstract. Artificial overwintering habitats for polyphagous predators. In: Watt TA, Buckley GP, eds. *Hedgerow management and nature conservation.* Wye College Press, Wye College, University of London, 157-58.

Marshall JK. (1967) The effect of shelter on the productivity of grasslands and field crops. *Field Crop Abstracts* **20**, 1-14.

Moore NW, Hooper MD, Davis BNK. (1967) Hedges. I. Introduction and reconnaissance studies. *Journal of Applied Ecology* **4**, 201-20.

Murton RK, Westwood NJ. (1974) Some effects of agricultural change on the English avifauna. *British Birds* **67**, 41-69.

O'Connor RJ, Shrubb M. (1986) *Farming and birds.* Cambridge: Cambridge University Press.

Osborne PJ. (1982) *The effects of Dutch Elm disease on farmland bird populations.* Unpublished DPhil Thesis. Oxford.

Osborne PJ. (1984) Bird numbers and habitat characteristics in farmland hedgerows. *Journal of Applied Ecology* **21**, 63-82.

Parish T, Sparks TH, Lakhani KH. (1994) Models relating bird species diversity and abundance to field boundary characteristics. In: Watt TA, Buckley GP, eds. *Hedgerow management and nature conservation.* Wye College Press, Wye College, University of London, 58-79.

Parslow JLF. (1969) Breeding birds of hedges. *Monks Wood Experimental Station Report (1966-68).* No. 21. Abbots Ripton, Huntingdonshire.

Pollard E. (1968) Hedges. II. The effect of removal of the bottom flora of a hawthorn hedge on the fauna of the hawthorn. *Journal of Applied Ecology* **5**, 109-23.

Pollard E. (1971) Hedges. VI. Habitat diversity and crop pests. *Journal of Applied Ecology* **8**, 751-80.

Pollard E. (1973) Hedges. VII. Woodland relic hedges in Huntingdonshire. *Journal of Ecology* **64**, 343-52.

Pollard E, Hooper MD, Moore NW. (1974) *Hedges.* London: Collins.

Pollard E, Relton J. (1970) Hedges. V. A study of small mammals in hedges and cultivated fields. *Journal of Applied Ecology* **7**, 549-57.

Roseman J. (1991) An archaeological study of field and parish boundaries. *Wiltshire Archaeological and Natural History Magazine* **84**, 71-83.

Rothery F. (1994) Abstract. Management of hedgerow vegetation for weed control and enhancement of beneficial insects. In: Watt TA, Buckley GP, eds. *Hedgerow management and nature conservation.* Wye College Press, Wye College, University of London, 159.

Stiva (1800). Strictures on hedge management. *Farmers' Magazine* **1,** 371-73.

Turner JH. (1845) On the necessity for the reduction or abolition of hedges. *Journal of the Royal Agricultural Society* **6,** 479-88.

Williamson K. (1969) Habitat preferences of the wren in English farmland. *Bird Study* **18,** 81-96.

CHAPTER 2

HEDGEROW MANAGEMENT: PAST AND PRESENT

N.R. Bannister[1] and T.A. Watt[2]
[1] *Brenoweth, Grampound, Truro, Cornwall TR2 4RB*
[2] *Wye College, University of London, Wye, Ashford, Kent TN25 5AH*

Introduction

Hedgerows are the oldest functioning man-made feature in the landscape (English Heritage, 1991) and there is increasing awareness of their age. Over the past few decades numerous papers, pamphlets and books have been written about hedgerows; their history, wildlife and management (Pollard *et al.*, 1974; Muir and Muir, 1987; Dowdeswell, 1987; Bannister, 1991; Maclean, 1992; Clements and Tofts, 1993). The recent resurgence in literature has been stimulated by the Countryside Commission's Hedgerow Incentive Scheme (HIS), now in its second year (Whelon, 1994, this volume) and the recent Private Member's Bill on hedgerow protection (Wilson, 1994, this volume). The importance of hedgerows for wildlife, for their history, for their landscape value and for various agricultural activities is now widely acknowledged.

The emphasis of the HIS and other conservation schemes such as Countryside Stewardship and Environmentally Sensitive Areas is on the resumption or continuation of traditional management techniques. Hedgerow management evolved from a basic premise that repeated cutting of woody shrubs stimulates new growth, resulting in a bushy structure compared with an unmanaged tree. The technique of manipulating woody species to harvest a crop and/or to encourage new growth has been used since early prehistory. A possible scenario is that Mesolithic people, in making clearings in the wild wood to trap grazing herbivores, found that straight, clean poles grew from the stumps which could be used both for tools and in the construction of shelters. Archaeological evidence from the Somerset levels has shown that in the Neolithic period, farmers were actively coppicing wood to make hurdles for trackways (Coles *et al.*, 1973; Coles and Orme, 1976). It is only a small step for hurdles to be used as 'dead' hedges to protect crops or to enclose animals.

Early literature

The first written evidence for the practice of laying hedges comes from Caesar's campaigns in Gaul. Here the Nervii people might, 'the more easily entangle their neighbour's horse, if they made excursions into their territories to plunder; splitting young trees down the middle and bending them down filling up the middle with brambles and thorns, they formed these hedges into fortifications like unto a wall; through which it was not only impossible to penetrate but also to be discovered.'   *Commenteria de Bello Gallico c.* 55 BC (Towers, 1755)

Hedges were a feature of the English landscape in the Anglo-Saxon period and many boundaries were already old at the time of the writing of the Saxon Charters (Rackham, 1986). Master Fitzherbert, writing in the sixteenth century, describes the 'plashing or pleching' of a hedge, the method of which had altered little since Caesar's campaigns.

'if the hedge be .x or .xii yeres growing sythe it was first sett, thanne take a sharped nachet or handbyll and cutte the settes in a playne place, nighe vnto the erthe, the more halve a-sander; and bende it doune toward the erthe, and wrappe and wynde theym together ...' (Fitzherbert, 1534)

It is very probable that even in Fitzherbert's time, regional variations in hedge management had evolved and were continuing to alter, reflecting the different functions of the hedge and local traditions.

The Agricultural Revolution

The rapid changes in agricultural methods in the eighteenth and nineteenth centuries could only succeed with effective enclosure of the open fields and the improved management of the existing boundaries. At the same time the new methods in husbandry were being recorded by landowners in surveys and commentaries. In the late eighteenth century the Board of Agriculture and Internal Improvement commissioned a series of General Views or Reports on the Agriculture in England on a district basis. These give valuable observations on regional differences in boundary management; for example, methods of planting, rejuvenation of old hedges, and management of existing ones. The General Views recorded not only the widely accepted methods of management, but also commented on 'experiments' undertaken by respected landowners. A number of the reports looked at the current state of enclosure within the county or district, often recommending areas which would benefit from enclosing (Middleton, 1807; Stevenson, 1809; Vancouver, 1810).

The methods for management varied depending on the age of the hedge, its present state of growth and its function in the landscape.

Management of young hedgerows

The trimming of newly planted hawthorn hedgerows was not always advocated by landowners. The first 'plashing' or laying after planting in 'kindly soil' took place after five years of unhindered growth (Banister, 1799). At the same time the ditch was scoured out and the spoil placed on the roots of the quicks.

For those hedgerows planted on soils 'less kindly in their propagation' and which were stunted and weak, cutting hard back in the fourth or fifth year encouraged the plant to 'fling out fresh shoots'. This sequence was repeated several times and improved many poor hedges (Banister, 1799). Anderson (1784) observed that the cutting and shortening of top or leading shoots also weakened a hedge. He argued that the strength of a hedge lay in it having 'proper, upright stems' which could stop an 'angry bull'. The strength did not come from numerous shoots. To achieve strong stems young hedgerows were allowed to grow vertically unhindered save for careful pruning of any weak top shoots; side shoots were thinned to avoid dense shading. Once the desired height was achieved, the side shoots were trimmed very close to the main stem for a couple of years, to encourage a flush of growth at the base. The young hedge was then left to grow freely (Anderson, 1784).

A method favoured by one Dorset farmer establishing a hedge in an exposed situation, was to plant three rows of quicks of four or five years old. These were kept low by continual clipping. 'He (Mr Bridges) has some good fences on high land, which were planted twenty years ago' (Stevenson, 1815).

The general practice advocated eighty years later was to cut a new hedge back to a stump of three or four inches after two to three years. Then to trim to a hog-maned triangular shape to minimize shading of lower shoots by those above. The hedge was then allowed to grow up to six feet tall and 'wattled' (laid) at an angle of forty degrees (Malden, 1899).

Traditional management of established hedgerows

Poor but healthy hedges evidently responded well to plashing or 'the wattling of living wood' (Anderson, 1784; Malden, 1899).

However, not all hedges were laid or 'plasshed' Some, as in Norfolk, were coppiced to ground level, and protected from grazing stock with dead hedges or hurdles (Marshall, 1767). In Cheshire, it was the custom instead of plashing, to renew a hedge by cutting the stems to about two feet of the ground, with the smaller parts of the branches run between the stumps. The subsequent regrowth was clipped with shears giving a pleasing effect (Wedge, 1794). This method was also seen as leading to a decline in hedge quality. Cutting old hedges back to two foot high ('buckstalling') and scouring the ditch for silts without replacing any on the bank or roots of the hedge ('out holling'), resulted in 'destitute hedges which dwindled away, stub after stub until they were no longer fences' (Marshall, 1767).

The frequency and timing of cutting hedgerows varied around the country with some advocating that cutting every year was detrimental to the growth of the hedgerow. Frequent cutting of,

'quicks or other plants keeping them small and weak in the stems, and rendering the hedges as they grow old, open at the bottoms' (Dickson, 1804).

He suggested that,

'those hedges cut at proper intervals, as about every seven or eight years, have a general superiority over those which have been clipped from the time they were first planted' (Dickson, 1804).

Malden (1899) quotes a maxim that 'a hedge should be trimmed whenever the knife is sharp enough' and that a summer trim controls the growth of weeds and thus ensures encouragement of the bottom growth of the hedge.

Frequent clipping with shears can raise a good thorn hedge, with continual clipping as the shoots advance. This gives the hedge a 'neat and pleasing aspect' (Banister, 1799). The trimming is best achieved, according to Malden (1899), by using a plashing bill with an upward stroke.

It was agreed that hedges should be clipped only when the adjacent fields were planted to wheat the following season, or to place hurdles along newly cut hedges in pasture fields, to avoid stock grazing (Anderson, 1784; Banister, 1799; Malden, 1899).

Traditional hedgerow management was labour intensive, but the hedge and its products were seen as a valuable resource and asset to the land holding. The farmer could with careful management procure,

'hurdle rods, wythes, poles and other useful stuff from the spontaneous vegetation of the ash, elm and other saplings ... supplying him with firewood etc. free of charge' (Banister, 1799).

Writing in the late eighteenth century, Banister observed that it was common practice in this country to leave twenty feet as a headland next to the hedge for the growth of a grass sward 'providing a quantity of herbage'. He also suggested that it was customary to plant fruit trees along this headland from which farmers could derive additional income from the fruit (Banister, 1799).

Despite the variations in management one thing was commonly agreed. The frequently preferred hedging plant was white thorn or 'quickthorn' namely *Crataegus monogyna* (common hawthorn).

'It possesses in a more eminent degree than any other plant common, with us, the requisite qualities of quickness of growth, strength, prickliness, durability and beauty' (Anderson, 1784).

Vernon (1899) recommended hawthorn for its quick growth, long life, rigid habit, and stubborn character.

Management of old and 'relict' hedgerows

Coppicing was favoured as a means of improving old hedgerows. Diseased hedgerows would be cleaned of all dead material and the remaining stems cut back to within two inches of the ground (Anderson, 1784) This occurred in the winter with the subsequent removal of all but a few buds the following spring, which then grew into strong shoots. However, it was found that repeated coppicing only produced numerous weak uprights. Anderson (1784) also advised that fallowing the adjacent fields and applying a rich dressing of manure along the margins encouraged 'proper' growth of a recently coppiced hedgerow.

At about the same time it was observed in Norfolk that there was an 'unpardonable custom of taking off the side shoots of tall hedges, leaving the top shoots to overhang young shoots and shade out their growth' (Marshall, 1767).

Twentieth century management

Since the beginning of this century there has been a decline in hedgerow management with respect to both the quality (the ability to restrain stock) and quantity (total lengths). This has been brought about by increased mechanization, the convenience and economics of using a flail hedge cutter, the intensification of arable farming and the cost of labour (Sturrock and Cathie, 1980).

For example, in a study of agricultural landscapes, many hedges in Huntingdonshire, already of poor quality in 1974, had disappeared by 1984 (Countryside Commission, 1984). This was attributed to 'intensive arable cultivations, continual hard trimming, plus possible occasional herbicide treatment and burning'. It was also found that 'lack of maintenance led to a general decline in hedge quality' and that *C. monogyna* was not suited to mechanical cutting in some (unspecified) conditions.

More recent surveys of British field boundaries have shown that twenty per cent of boundaries classified as hedgerows in 1984 were classified as a different type of boundary, for example, lines of trees or shrubs or relict hedgerows, when re-surveyed in 1990; thus suggesting that less active management has taken place during the 1980s (Barr *et al.*, 1991).

Recent historical research has shown that, in Kent, many wide hedgerows or 'shaws' were either being cut back hard to form a thin line of shrubs or were removed altogether (Bannister and Bannister, 1993*a,b*).

This practice had already been noted in the eighteenth century;

'indeed a spirit for clearing away crooked wide hedgerows and raising straight thin hedges in their steads may be said to be now high in the using among the yeomanry of this part (Maidstone) of Kent' (Marshall, 1798).

Recent studies on hedgerow growth

The change from intensive, manual management of hedgerows to either intensive mechanized trimming or a complete absence of management has affected the quality of England's hedgerow stock. Despite the many surveys and reports on hedgerows, there has been little work specifically on the management of hawthorn (*Crataegus monogyna*).

Experiment

A recent MAFF-funded research project at Wye College investigated the growth response of a newly planted hawthorn hedge to different cutting regimes (Bannister, 1991). The hawthorn slips were planted following guidelines given by the Ministry of Agriculture, Fisheries and Food (ADAS, 1980) and the effects of both the season and position of cutting were examined under controlled conditions.

The four types of cutting, carried out in either winter (February) or summer (August) were;

i. a vertical trim which removed lateral shoots over 10 cm in length back to 10 cm from the main stem;
ii. a horizontal trim where all the leading shoots were removed to a height of 70 cm above the ground;
iii. a combined horizontal and vertical trim;
iv. no management.

Hawthorn was found to respond to trimming in similar ways to other woody shrubs. If the leading shoots were removed (i.e. a horizontal cut), fewer but longer shoots were produced the following year which tended to end in an apical bud rather than a thorn. If a horizontal cut was combined with a vertical cut, which also removed the lateral shoots, apical bud formation was also enhanced. These responses were similar to those obtained by pruning apple trees (Mika, 1982, 1986).

When only a vertical cut was applied to the plants, a summer cut stimulated thorn-tipped shoot production compared to a corresponding cut in winter. A combined vertical and horizontal cut stimulated a greater response in shoot numbers than a corresponding winter cut. Summer cutting took place when apical dominance was weak, resulting in a flush of shoots the following year. Winter cutting produced fewer shoots, but these tended to be longer and more vigorous.

This work has management implications for hedgerows: if a bushy hedge is required, trimming should be in summer so as to produce many new shoots the following year. If, however, a strong upright hedge is required a horizontal cut in winter should give the desired effect with long, vigorous leading shoots.

Survey

The growth of hawthorn was also observed in field hedges under different management regimes to see if their effects could be identified by different growth characteristics (Bannister, 1991).

Hawthorn-dominated hedges in Kent and Leicestershire were selected for measurement. Their management ranged from annual flailing to hand-cutting, including both winter and summer cutting. Hedgerows were grouped successfully by management types according to growth characteristics using Canonical Discriminant Analysis (SAS Institute Inc, 1985).

A handcut hedge was characterized by numerous, short shoots which tended to end in an apical bud, therefore, this management resulted in a thick, dense hedge with all the growth on its outer edges, and leads us to speculate that before mechanization, all trimmed hedges had a far denser appearance than they do today.

It was found that winter cutting and unmanaged sites had a larger leaf area per square metre than summer-cut hedgerows. However, a summer cut was also characterized by the presence of *Galium aparine* which may have had a smothering effect on the summer-trimmed hedgerows. Die-back of hawthorn hedgerows caused by smothering by bracken (*Pteridium aquilinum*) has been observed in Essex (N.R.Bannister, personal observation). The use of herbicide sprays in hedgerow bottoms to kill noxious weeds results in the removal of perennial grasses and allows persistent weeds, such as *G. aparine* to become dominant (Froud-Williams *et al.*, 1980). The long-term effect of weed infestation on hedgerows is not known, but it could be suggested that one reason for the decline in quality of many arable hedgerows is partly due to the suppression of shoot and leaf growth and thus carbohydrate production by the smothering effects of climbing weeds. This, combined with the cessation in laying which cleaned a hedge of dead wood and weeds, could be the main cause of poor, gappy hedgerows in arable situations. Intense stock grazing of unfenced hedgerows in pasture fields also leads to gappy hedgerows.

In the field survey, a hedge of over two metres high which was severely cut back using a flail, was observed to have very little new shoot growth the following season, but eighteen months after treatment it had a very vigorous growth. The use of the flail on the current season's growth of shoots did not appear to have a detrimental effect on the 'production' of new shoots the following season. This is in keeping with the findings of a recent research project which examined the effect of different intensities of flail cut on hawthorn (D. Semple, personal communication).

Conclusions

The neglect of hedgerows was recognized over a century ago (Malden, 1899). Since then, changes in methods of management and the redundancy of hedgerows as a functional element in some agricultural systems have meant that the hedgerow resource in England has declined in both quantity and quality. The various Government schemes recently implemented to reintroduce traditional hedgerow management, are an attempt to redress the situation. Our knowledge of the subtleties of the differences in regional techniques of management and the response of the hedge to different methods of trimming is still very limited.

References

ADAS (1980) *Planting farm hedges.* Leaflet No. **763**. London: Ministry of Agriculture, Fisheries and Food. 11pp.

Anderson J. (1784) *Essays relating to agriculture and rural affairs.* Edinburgh: John Bell, page 20.

Banister J. (1799) *Synopsis of husbandry: being cursory observations in the several branches of the rural economy adduced from a long and practical experience in a farm of considerable extent.* London: G.G. & J. Robinson, page 452.

Bannister NR. (1991) The management and growth of hawthorn (*Crataegus monogyna* Jacq.) under experimental conditions and on Lowland farms in England. Unpublished PhD Thesis, University of London, 16-59.

Bannister NR, Bannister DE. (1993*a*) *The historic landscape survey of Great Tong Farm, Headcorn, Kent.* Unpublished Report for English Heritage, Archaeology Division, Fortress House, 23 Savile Row, London WIX 1AB. 198pp.

Bannister NR, Bannister DE. (1993*b*) *The historic landscape survey of Wormshill Estate, Sittingbourne, Kent.* Unpublished Report for English Heritage, Archaeology Division, Fortress House, 23 Savile Row, London WIX 1AB. 183pp.

Barr C, Howard D, Bunce R, Gillespie M, Hallam C. (1991) *Changes in hedgerows in Britain between 1984 and 1990.* Report, Institute of Terrestrial Ecology, Merlewood.

Clements DK, Tofts RS. (1993) Hedges make the grade - a look at the wildlife value of hedges. *British Wildlife* **4,** 87-95.

Coles JM, Hibbert FA, Orme BJ. (1973) Prehistoric roads and tracks in Somerset: 3. The Sweet Track. *Proceedings of the Prehistorical Society* **39,** 256-93.

Coles JM, Orme BJ. (1976) A neolithic hurdle from the Somerset Levels. *Antiquity* **50,** 57-61.

Countryside Commission (1984) *Agricultural landscapes: a second look.* Cheltenham, CCP 168, page 20.

Dickson RW. (1804) *Dickson's agriculture,* Vol. 1. London: Richard Philips, page 130.

Dowdeswell WH. (1987) *Hedgerows and verges.* London: Allen and Unwin. 190pp.

English Heritage (1991) *Farming historic landscapes and people.* London: English Heritage, 16pp.

Fitzherbert, Master (1534) Book of husbandry. Reprinted from the 1534 edition by Revd. Walter W. Skeat (1882). London: English Dialect Society, page 78.

Froud-Williams RJ, Pollard F, Richardson WG. (1980) Barren Brome: A threat to winter cereals? *Report of the Weed Research Organisation,* 1978-1979, 43-51.

Maclean M. (1992) *New hedges for the countryside.* Ipswich Farming Press. 276pp.

Malden WJ. (1899) Hedges and hedge-making. *Journal of the Royal Agricultural Society* **10,** 87-117.

Marshall TW. (1767) *The rural economy of Norfolk comprising the management of landed estates.* London: T. Cadell, page 96.

Marshall TW. (1768) *Rural economy of the Southern Counties, The Eastern Division of Chalk Hills, London.* **2,** page 34.

Middleton J. (1807) *View of the agriculture of Middlesex*. Drawn up for the Board of Agriculture and Internal Improvement, London, page 144.

Mika A. (1982) The relation between the amount and type of pruning and the yield of apple trees. *Proceedings of the 21st International Horticultural Congress* **1,** 209-21.

Mika A. (1986) Physiological responses of fruit trees to pruning. *Horticultural Review* **8,** 337-78.

Muir R, Muir N. (1987) *Hedgerows: their history and wildlife*. London: Michael Joseph. 250pp.

Pollard E, Hooper MD, Moore NW. (1974) *Hedges*. New Naturalist Series No. **58.** London: Collins. 256pp.

Rackham O. (1986) *The history of the countryside*. London: J.M. Dent & Sons Ltd, page 185.

SAS Institute Inc. (1985) *Users guide: statistics version,* 5th edn. Cary NC: SAS Institute Inc. 956pp.

Stevenson W. (1809) *General view of the agriculture of the county of Surrey*. London: Richard Phillips, page 143.

Stevenson W. (1815) *General view of the agriculture of the county of Dorset*. London: Sherwood, Neely and Jones, page 175.

Sturrock FG, Cathie J. (1980) *Farm modernization and the countryside: the impact of increasing field size and hedge removal on farms*. Department of Land Economy, University of Cambridge, Occasional Paper No. **12.** 69pp.

Towers J. (1755) *Caesar's commentaries of his war in Gaul* (with an English translation). London: Hutch and Hawes, page 67.

Vancouver C. (1810) *A general view of the agriculture of Hampshire including the Isle of Wight*. Drawn up for the Board of Agriculture and Internal Improvement, London, page 121.

Vernon A. (1899) *Estate fences: their choice, construction and cost*. London: E & F.N. Spon Ltd, page 30.

Wedge T. (1794) *General view of the agriculture of the County Palentine of Chester*. Board of Agriculture and Internal Improvement, London, page 30.

Whelon D. (1994) The Hedgerow Incentive Scheme. In: Watt TA, Buckley GP, eds. *Hedgerow management and nature conservation*. Wye College Press, Wye College, University of London, 137-45.

Wilson A. (1994) The fight for hedgerow protection legislation. In: Watt TA, Buckley GP, eds. *Hedgerow management and nature conservation*. Wye College Press, Wye College, University of London, 146-49.

CHAPTER 3

RENOVATION AND EXPLOITATION OF HEDGES AROUND GRASSLAND

E. J. Asteraki[1], R. O. Clements[1], G. O'Donovan[2], B. C. Clifford[2], A. T. Jones[2,] R. J. Haggar[2] and B. J. Thomas[2]

[1] *Institute of Grassland and Environmental Research, North Wyke, Okehampton, Devon EX20 2SB, UK*

[2] *Institute of Grassland and Environmental Research, Plas Gogerddan, Aberystwyth SY23 3EB, Wales, UK*

Introduction

Hedges are an important component of the grassland ecosystem. Initially, of course, they were established as stock-proof barriers between fields or farming units and clearly many still serve that same purpose. However, they have many other functions, including becoming associated with an attractive and natural countryside and providing a refuge or habitat for a wide range of plants and animals. It is unfortunate then that many hedgerows have become derelict and no longer serve their basic function, let alone retained an attractive appearance or their usefulness as a wildlife habitat.

Many of the species of flora and fauna associated with hedgerows have little direct impact on man's farming activities or general welfare. However, some species, including the group of predatory beetles, Carabidae, are directly beneficial and others, including some plant disease organisms, are deleterious. The present paper outlines three separate but cognate topics of work which examined (1) the feasibility of re-constructing derelict hedgerows and how this process could be accelerated, (2) the impact of some hedgerow management techniques on the beneficial carabid fauna and (3) an attempt to monitor the movement of certain virus diseases out of a hedgerow into the clover component of an adjacent sward.

Hedge replanting

In many hedges in western Britain, dereliction has proceeded to the point where some replanting is required. Dereliction was estimated during 1976-79 to occur in 46% of a large sample of hedges in mid-Wales (Jones *et al.*, in press). Dereliction can range from small gaps in hedges to whole lengths where shrubs are either absent or are in a moribund state. The most appropriate replanting sites are often on the tops of boundary banks. This is because boundary banks are the commonest field boundary, accounting for 98% of hedge sites in the above survey. It was anticipated that it would be difficult to establish shrubs upon banks because they are free-draining, lack deep soil and the crests are very exposed.

Materials and methods

It was decided to test the potential of soil nutrient applications and a peat ameliorant to aid shrub establishment in 'difficult' bank-top sites. The two most frequently planted shrub species *Crataegus monogyna* (hawthorn) and *Prunus spinosa* (blackthorn) were compared for this purpose at two upland sites (*c*. 300 m in altitude), Pwllpeiran and Bronydd Mawr, with harsh, exposed climates and earth/stone banks to provide a 'worst scenario'. At Pwllpeiran, 20 km east of Aberystwyth, the bank was about 0.5 m high with a flat top but with rubble infill; at Bronydd Mawr, near Brecon the bank was mainly of earth infill, at least 1.5 m high with a 0.5 m wide crest and steep sides (>45°).

To compare the effect on survival of the two shrub species of fertilizer applications and a mulch ameliorant, five treatments were set up at Bronydd: NPK, P only, peat only, NPK + peat, control (no inputs); and three at Pwllpeiran: NPK, P only, control (no inputs). The experimental layout was such that at each site there were four replicates for each species/treatment combination arranged in four blocks. Each replicate consisted of a 2.3 m length of bank crest planted in a zig-zag double-row fashion with 21 whips, following ADAS recommendations (ADAS, 1986). Sites were fenced on each side, at least 2 m from the hedgerow, to protect shrubs from sheep grazing. Shrubs were planted in early March 1990, as 45-60 cm whips for hawthorn and 40-60 cm whips for blackthorn. Shrubs were planted in slits deep enough for the roots to spread without constraint and then heeled in.

There were no recommendations for fertilizer rates for shrubs found in the literature, though there was some information on the rates for trees (Davies, 1987; Taylor, 1991), so rates were applied which were equivalent to heavy grassland rates to 'exaggerate' any effect. The rates per plant, where applied, were 11 g of triple superphosphate (46% P_2O_5) and 20 g of 'kay-nitro' (25-0-16) to provide phosphorus, nitrogen and potassium, the granular fertilizers being spread evenly over a 0.25 m^2 quadrat centred over each shrub giving a total equivalent to 200 kg N, 88 kg P and 128 kg K/ha. At Pwllpeiran, the experiment also included the application of lime. Where peat was used, it was moistened and one litre per shrub was packed around the roots prior to heeling in. Survival was measured in September - prior to leaf fall.

Results

At Bronydd, mean survival was high (73%). There were, however, highly significant differences in survival, both between species and between treatments (Table 3.1A). Hawthorn showed overall a higher survival than blackthorn (84% compared with 68%). The NPK treatment depressed the percentage survival of both species, though this effect was more pronounced in blackthorn. The P only, peat only and control treatments showed similar survival values which were higher than NPK treatments, both with and without peat.

For the Pwllpeiran experiment (Table 3.1B), results were similar to those found at Bronydd, the overall survival being high at 81%. Hawthorn again showed a higher overall percentage survival than blackthorn (87% compared with 75%) and NPK

Table 3.1 (A) Mean survival at Bronydd Mawr

| Species | Treatments | | | | | |
	NPK	P	Peat	NPK + peat	Control	Overall
Hawthorn	67	98	98	60	98	84
Blackthorn	31	98	88	36	86	68

ANOVA: Treatments *F* Prob. <0.001; Species *F* Prob. <0.001; Interaction *F* Prob. = 0.08.

Table 3.1 (B) Mean survival at Pwllpeiran

| Species | Treatments | | | |
	NPK	P	Control	Overall
Hawthorn	81	90	92	87
Blackthorn	52	87	86	75

ANOVA: Treatments *F* Prob. <0.003; Species *F* Prob. <0.02; Interaction *F* Prob. = 0.09.

treatments depressed the percentage survival of both species, with a greater effect in blackthorn.

Discussion

Far from aiding establishment of the shrub species, the addition of NPK fertilizer gave rise to higher death rates than the other treatments. This finding bears comparison with that of Davies (1987) who found that survival and growth of ash, Norway maple and Norway spruce, planted to subsoil, may be reduced by applications of NPK fertilizer. Two factors which may have contributed to the higher death rate in the present study are fertilizer burn and competition from grass.

There was a pronounced drought in the spring of 1990 with no significant falls of rain until June, followed by more dry weather. It is possible that the 'undiluted' fertilizer application caused some burn to the shrub roots. Small patches of turf were killed where pockets of fertilizer granules collected. Though quantitative estimates were not made, grass production in plots receiving NPK was at least twice as great as other plots by the end of the growing season, despite the early drought, and the increased growth could have resulted in higher levels of competition. This grass growth was surprising given the dryness of the soil at the top of the bank. Competition for water in such dense vegetation could be critical in such a free draining and droughted site.

The evidence suggests that fertilizers have no role to play in aiding establishment of newly-planted hedges on bank-top sites as they exacerbate the problem of competition from grass. Prior to this experiment, grass competition was only conceived to be a problem in establishing new hedges on flat, fertile land where many potential sites have received considerable inputs of fertilizer (e.g. other hedges planted at Pwllpeiran and Bronydd).

The greater overall survival of hawthorn over blackthorn is surprising as blackthorn whips have much more extensive roots (at least the length of the stem) than hawthorn (about one-third of the stem length). The hawthorn whips have an extremely corky bark compared with the smooth bark of the blackthorn and this may aid water conservation. The greatest difference in survival rates between species was manifest in the NPK treatment with blackthorn showing approximately twice the mortality of hawthorn. In reducing survival of the two species either or both nitrogen and potassium may be detrimental as phosphorus alone did cause such high mortality.

Bearing in mind the spring drought at both sites, the most sensible time for planting banks would be early autumn, so that roots can establish in warm, damp soils before the onset of summer drought.

Beneficial invertebrates

Spraying grassland with the most commonly used insecticide on pasture (chlorpyrifos) had a deleterious effect on the number and diversity of carabids caught (Asteraki *et al.*, 1992*a*). Their populations recovered after a year, probably as a result of their re-invasion from surrounding hedgerows. Pollard (1968) showed that hedgerows also contribute to the diversity of carabid species present in cultivated areas and it is thought that field boundaries around cereals provide an important overwintering site and source of potential beneficial predatory activity (Sotherton, 1985). It is important, therefore, to study the carabid fauna of hedgerows around grassland and to assess the impact of various management practices, including herbicide use.

Materials and methods

A semi-permanent pasture at Hurley with a well-established hawthorn hedge was selected (Asteraki *et al.*, 1992*b*). The hedge and field margin were divided into nine plots, each 15 m wide and 45 m long. In August 1988 the plots were separated by polythene barriers, inserted 50 cm into the soil with 50 cm remaining above ground, to prevent movement of the beetles from one plot to another. A 2 m strip at the base of the hedge remained uncut whilst the remainder of the field was mown to keep sward height between 6-12 cm.

After inserting the barriers, but prior to the application of treatments, a baseline study was implemented to assess the diversity and distribution of the existing carabid and spider fauna.   In May 1989, and again in September 1989, experimental treatments were applied to the strips at the base of the hedge.   The nine plots were divided into three replicate blocks, the treatments being assigned randomly within the blocks.   The treatments were; R: herbicide to kill all flora at the base of the hedge (glyphosate; 5 litres 'Roundup'/ha in 300 litres water); B: herbicide to kill all broadleaved weeds (2, 4-D + dicamba + triclopyr; 4 litres 'Broadshot'/ha in 400 litres water); C: control (untreated).

Carabids were sampled using pitfall traps (Clements *et al.*, 1988), with ethylene glycol as a preservative. Pitfall traps were located at 2, 5, 10, 20, and 40 m from the hedge. There were two traps at each distance in each plot. Samples were collected fortnightly, from August 1988 through April 1989 before treatment application and from May 1989 through November 1989 after treatment.   To reduce inherent variation in the data they were transformed to percentages of the total, thus eliminating the differences in total catch. In order to assess the impact of the removal of the flora at the hedge bottom the data were subjected to detrended correspondence analysis, DECORANA (Hill, 1979). This program ordinates sites and species on four axes, each axis being a measure of variation in declining order of significance.

Results and discussion

Two weeks after treating the bottom of the hedge, marked differences in vegetation could be observed between plots. The 'Roundup' treatment killed all of the flora and the 'Broadshot' treatment all of the broadleaved species (Table 3.2).

Table 3.2   Flora found at the base of the hedge in each treatment

Treatment	Flora of field margin
'Broadshot'	*Poa annua, Lolium perenne, Elymus repens, Holcus lanatus*
'Roundup'	No living flora, bare soil
'Control	*Poa annua, Lolium perenne, Elymus repens, Holcus lanatus, Galium aparine, Anthriscus sylvestris, Lamium album, Urtica dioica, Stellaria media, Capsella bursa-pastoris, Papaver dubium, Veronica hederifolia, Alliaria petiolata, Cirsium arvense, Arum maculatum*

A DECORANA plot for carabids caught before treatment showed no separation between the plots of each treatment group. Consequently, any differences observed after treatment application were due to the herbicides and not to any inherent variation between the plots. Figure 3.1 gives the DECORANA plot for carabids caught after treatment using the data for 2, 5 and

10 m from the hedge, a clear separation can be observed along axis 2, indicating a significant effect of the herbicide treatment.

Figure 3.1 DECORANA ordination plot of carabid percentages after treatment application, showing treatment groups joined to form polygons, representing 2, 5 and 10 m sampling distances from the hedge

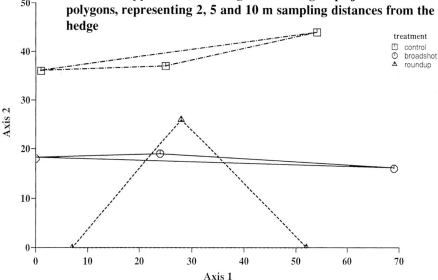

A DECORANA ordination for carabids caught before treatment, showed large separation along axis 1 between beetles caught at 2 m from the hedge and those caught at the other distances and this separation persisted after treatments were applied. It was clear from the DECORANA plots that there was a significant difference between the carabid community in the hedge and that as little as 10 m away. From the ranking produced in the DECORANA output this appeared to be due to species such as *Synuchus nivalis, Pterostichus strenuus, Agonum dorsale,* and *Leistus ferrugineus.* All these species are regarded as woodland fauna (Thiele, 1977), the hedge itself providing a suitable habitat in which these species can survive. *A. dorsale* is an exception to this, migrating from the field to the hedge in the winter (Pollard, 1968).

The herbicide treatments affected five key species, *S. nivalis, Harpalus rufipes, Amara aenea, Amara plebeja,* and *Loricera pilicornis. S. nivalis,* as previously explained, is a woodland species, living in the damp moss and litter layer of the woodland floor. This litter layer was removed by the 'Roundup' treatment, thus depriving this species of a suitable habitat. *H. rufipes, A. aenea* and *A. plebeja* are all omnivorous species, plant material and seeds making up a large part of their diet. The herbicide treatments both reduced the amount and diversity of plant material and the seeds available. The ideal habitat for *L. pilicornis* is damp soil, but in the 'Roundup' treatment the soil was devoid of all plant cover and exposed to the particularly warm weather experienced during that season, causing the soil to become dry and hard.

The effect of altering the flora at the base of a hedge could be two-fold: it could either deprive a particular species of suitable food, or it could change the habitat to such an extent that it would no longer survive in those particular surroundings. This work demonstrated that carabids are affected by herbicides applied to hedgerows, but more research is needed to discover the mechanisms of this effect.

Pests and pathogens

Viruses may be important pathogens of white clover (*Trifolium repens*), a key component of many swards. Their distribution in the UK and the importance of hedgerows as reservoirs of infection, however, is poorly understood. The present work surveyed the presence of two viruses and studied features that may affect their spread from hedgerows. A survey of virus infestation in pasture and lawns in Wales and parts of England was carried out. Two of the viruses found, white clover mosaic potexvirus (WCMV) which is mechanically transmitted and clover yellow vein potyvirus (CYVV) which is vectored by aphids were selected for further studies.

Materials and methods

To investigate the dynamics of virus infection, a field margin was created in the form of a raised bank within a grass/clover ley. This was sown with local wild flower seed and white clover (*Trifolium repens* cv. S184) in the spring of 1990. The clover plants were inoculated with WCMV and CYVV and several management regimes were imposed, namely grazing by sheep and cutting to simulate hay and silage harvests on the adjacent sward. Monitoring for virus vectors, especially aphids, was carried out on the clover component of the swards.

Related glasshouse experiments (i) monitored movement of virus within the clover plant and observed the effect of temperature on this movement; (ii) examined movement of virus between clover plants by simulating cutting and trampling; (iii) investigated the host range of the viruses by inoculating locally-abundant hedgerow legumes (*Vicia cracca, Vicia sativa, Lathyrus pratensis,* and *Lotus corniculatus*) which may act as alternative hosts in the field; (iv) investigated if virus-infected debris in the soil acts as a source of inoculum for newly-planted clover and (v) studied the effects of different cutting regimes on survival of virus in the plants.

Results and discussion

CYVV and WCMV were scarce in the pastures surveyed, but the latter was present in the lawns of staff working at IGER, where the virus is present in experimental plots. Given the infectivity of the virus and its apparent easy spread by Institute personnel it was surprising that the disease was not more widespread in pastures. CYVV was found in the field experiment in the sward adjacent to the hedge-bank studied, but did not persist for more than a year. WCMV did not spread from the hedge-bank, despite its high infectivity. In an agricultural context, the movement of CYVV/WCMV into and out of field margins seemed unimportant.

Glasshouse and growth room experiments to elucidate certain aspects of virus movement showed that the viruses do not reside in other related hedgerow legumes and are not readily passed from plant to plant by invertebrates that chew.

However, they are easily transmitted by mechanical means and increased temperature increased the speed at which virus moved within the plant. Under the conditions employed, the viruses did not survive in infected plant debris in the soil and infection of new seedlings by this means appears unlikely.

Conclusions

The three research programmes outlined above examined one approach of accelerating the re-establishment of derelict hedgerows, the effects of removing the hedge-bottom flora on predatory ground beetles and investigated the dynamics of virus infection that may originate in hedgerows.

Somewhat surprisingly, fertilizers, far from having a role in promoting the development of newly-planted hawthorn or blackthorn, seemed to reduce sapling establishment by encouraging weeds. Fertilizer 'burn' may also have been a factor.

The destruction of the flora at the base of a hedgerow using herbicides, greatly diminished the species diversity of predatory ground beetles. Hedgerows around grassland may not act as reservoirs of invertebrate predators to the same extent as those around cereal fields, but clearly diminution of their diversity should be avoided.

Virus infection, at least that by CYVV and WCMV did not seem important or widespread. The potential for their spread from hedgerows also appeared unimportant. However, these findings cannot be extrapolated to other diseases.

Acknowledgement

This work was supported by the Ministry of Agriculture, Fisheries and Food.

References

ADAS (1986) Hedge planting leaflet.
Asteraki EJ, Hanks CB, Clements RO. (1992*a*) The impact of two insecticides on predatory ground beetles (Carabidae) in newly-sown grass. *Annals of Applied Biology* **120,** 25-39.
Asteraki EJ, Hanks CB, Clements RO. (1992*b*) The impact of the chemical removal of the hedge-base flora on the community structure of carabid beetles

(Coleoptera, Carabidae) and spiders (Araneae) of the field and hedge bottom. *Journal of Applied Entomology* **113,** 398-406.

Clements RO, Asteraki EJ, Jackson CA. (1988) A method to study the effects of chlorpyrifos on predatory ground beetles in grassland. BCPC Monograph No. **40,** *Environmental effects of pesticides* 167-74.

Davies RJ. (1989) In: Patch D, ed. *Advances in practical arboriculture.* Forestry Commission Bulletin **65.** London: HMSO.

Hill MO. (1979) *DECORANA: a FORTRAN program for detrended correspondence analysis and reciprocal averaging.* Section of Ecology and Systematics, Cornell University, Ithaca, New York.

Jones AT, Chater AO, Evans P, Potter F, Taylor J. (in press) A survey of hedges and their woody species in Ceredigion, mid-Wales. *Journal of Applied Ecology.*

Pollard E. (1968) Hedges. IV. A comparison between the carabidae of hedge and field sites and area of a woodland glade. *Journal of Applied Ecology* **5,** 649-57.

Sotherton NW. (1985) The distribution and abundance of predatory Coleoptera overwintering in field boundaries. *Annals of Applied Biology* **106,** 17-21.

Thiele HU. (1977) *Carabid beetles and their environment.* Berlin: Springer-Verlag.

Taylor CMA. (1991) *Forest fertilization in Britain.* Forestry Commission Bulletin **95.** London: HMSO.

CHAPTER 4

THE EFFECT OF RESTORATION TECHNIQUES ON FLORA AND MICROFAUNA OF HAWTHORN-DOMINATED HEDGES

J. H. McAdam[1], A. C. Bell[2] and T. Henry[3]
Department of Agriculture for Northern Ireland, Newforge Lane, Belfast BT9 5PX
[1] *Agricultural Botany Research Division, Science Service*
[2] *Agricultural Zoology Research Division, Science Service*
[3] *Agricultural Technology Division, Greenmount College of Agriculture and Horticulture*

Introduction

Importance of hedges

In Northern Ireland 83% of the aggregate gross margins on all farms are attributable to grazing livestock and the average farm size is 35.1 hectares (Department of Agriculture for Northern Ireland, 1993). There are many small enclosed fields (mean size 1.8 hectares) and a large number of field boundaries. This results in a largely pastoral landscape where field margins are the principal determinable visual feature and there were estimated to be approximately 150,000 kilometres of hedgerows in 1979 (Graham, 1979). Total tree cover in Northern Ireland is 5.6 per cent, which is the lowest in the EU and it has been estimated (Webb, 1985) that hedgerows occupy an area three times that of deciduous woodland in Ireland and that 60% of broadleaved trees in Northern Ireland are to be found in hedgerows (Carlisle, 1990). Although Northern Ireland has a large number of hedges many are poorly maintained and managed and have become straggling and 'gappy'. Such hedges fail to provide stock-proof barriers and are of limited conservation value.

Loss of hedgerows

There is a paucity of information on the rate of hedgerow loss anywhere in Ireland but loss rates seem to be lower than those recorded for Great Britain where 28 per cent reduction in hedgerows was recorded between 1946 and 1975, though there is evidence that the rate of loss may now have slowed down and even been reversed in some areas (Soper and Carter, 1991). It has been estimated that in the Republic of Ireland just under 2 kilometres of hedge per square kilometre may have been lost between 1936 and 1982, an average total loss of just over 14 per cent (An Foras Forbartha, 1985). There is some evidence to suggest that most of this loss has occurred since 1973.

Northern Ireland currently has three Environmentally Sensitive Areas (ESAs) designed to help conserve areas of high landscape and/or wildlife value which are vulnerable to changes in farming practices.   Payments are offered to farmers willing to continue or adopt environmentally beneficial farming practices.   Within the ESAs a particular effort is being made to maintain and improve field boundaries (Department of Agriculture for Northern Ireland, 1992).   Within the Mournes, Antrim and Sperrins Areas of Outstanding Natural Beauty (AONB) and Fermanagh District, of an estimated 6.3 kilometres of hedgebank per square kilometre of land area, hedgerow loss was estimated at approximately 0.5 per cent per annum from Ordnance Survey Maps and with a total loss of 4 per cent between 1977 and 1986 (Cooper *et al.*, 1991).

In a field boundary survey of a complete river catchment in a traditional grassland farming area, carried out in 1991, Mallon (1992) sampled approximately 1000 hedges and concluded that the rate of boundary removal was approximately 1 per cent per annum (between 1975 and 1991). This is approximately twice the rate found by Cooper *et al.* (1991) for an adjacent AONB and three times the rate estimated for the Republic of Ireland (An Foras Forbartha, 1985).

Hedgerow management

Surgery and good management of hedges in poor condition are required if the potential benefits of field boundaries in poor condition are to be realized (DANI/DOE, 1987). Poorly managed hedges can be categorized as those suffering from neglect and which are overgrown, and those which have been over-managed and are losing their vigour.

If the stems of the hedge are too thick for laying (i.e. >100 millimetres thick at the base) they may be hard coppiced, i.e. cut as close to the ground level as possible. If the stump is healthy, new shoots will sprout just beneath the cut (DANI/DOE, 1987). Hellewell (1991) strongly discourages cutting an overgrown gappy hedge to 1.0-1.5 metres high because the stems will then grow 'bushy heads' and the base of the hedge will remain thin and not stockproof.

The older the hedge, the less likely it is to recover from hard coppicing. DANI/DOE (1987) suggest that old, rotting stumps should be removed. The soil in the gaps created should be removed and replaced with fresh soil into which young hawthorn plants may then be planted. After five to seven years the new growth from the coppiced stumps and from the newly planted quicks can be trimmed into shape. Thus coppicing rejuvenates old, healthy plants and provides an opportunity to gap up the hedge without restricting light to the new plants.

On arable farms in England, coppicing on an eight-to-twelve year rotation is an economic method of hedge management. However, since the predominant enterprise in Northern Ireland agriculture is livestock production, it is likely that coppicing will remain a seldom used method of hedge rejuvenation.

An alternative strategy for managing and restoring overgrown gappy hedges is laying. Laying involves partially severing the main stem of the plant close to ground level and then laying these at an angle of approximately 30 degrees. The main stems should be 3.0-3.5 metres high and about 50-100 millimetres thick at the base (Hellewell,

1991). Vertical stakes are driven into the ground along the length of the hedge to secure the laid stems. Willow or hazel binders can be woven along the top of the newly laid hedge. Coppicing and laying are carried out between mid-November and early March.

Provided the gaps are not excessively large, laying can provide an immediate stock-proof barrier. Other advantages over coppicing are that protective fencing may not be required and laying has an apparently less drastic effect on wildlife. Rands and Sotherton (1987) suggest that the replacement of laying by frequent mechanical trimming has reduced the amount of dead grass in hedge-bottoms adversely affecting nesting habitat quality for the grey partridge.

While hedge laying has a long tradition in many English counties (Brooks, 1988), it is rarely practised in Northern Ireland. Where the majority of the hawthorn stems are greater than 50-100 millimetres diameter at the base, it is questionable how feasible this management strategy would be.

Despite the wide range of advice being given on management of overgrown, 'gappy' hedges, little is based on scientific evidence and no proper comparison of the restoration techniques reviewed above has been carried out. Hence, an experiment to investigate the effect of a range of restoration techniques on flora and microfauna of hawthorn-dominated hedges was initiated at a number of sites in 1990 and 1991. Obviously this is a long-term experiment and some preliminary results are presented in this paper.

Materials and methods

In 1990 and 1991 a randomized block experiment was set up in 14 hedges at ten sites throughout Northern Ireland (Figure 4.1). Hedges were chosen according to the following criteria: they were hawthorn-dominant, had approximately 150 metres of uniform height and density of trees, were largely overgrown and unmanaged and both sides of the hedge were permanent pasture. Six sites were chosen from two Environmentally Sensitive Areas (where hedgerow restoration is particularly targeted) and eight from non-ESA areas. In each hedge, 25 metre lengths of the following treatments were imposed.

1. Unchanged control
2. Laid during late winter
3. Cut to 1.5 metres high with the stem bases nicked to encourage sprouting
4. Hard coppiced (to near ground level) with gaps planted with hawthorn plants
5. Hard coppiced with gaps planted with a mixture of species - blackthorn, hazel, holly, beech, and hawthorn

Hedges were fenced on both sides to exclude stock and were trimmed mechanically to approximately 1.5 metres every third year where appropriate.

Figure 4.1 Distribution of hedgerow restoration sites throughout N. Ireland

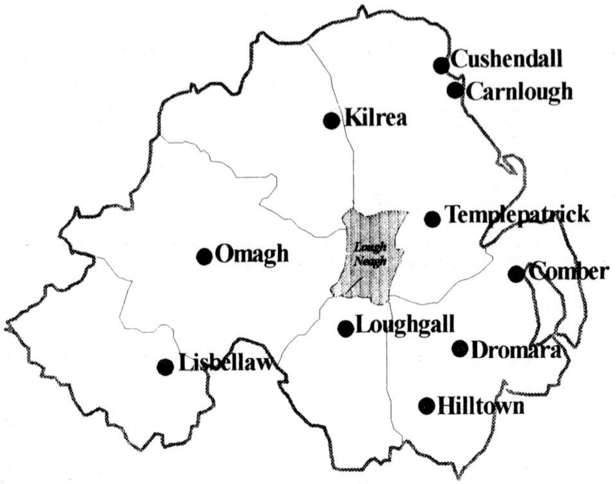

Recording

'Plots' were considered to represent 20 metre lengths of hedge, i.e. a 'guard' row of 10 metres of hedge existed between each plot. A complete plant species list was made for each plot (both sides of the hedge) and three permanent quadrats (1 metre x 1 metre x full height of hedge) were delimited on each side of the hedge. Within all six quadrats per plot, cover abundance estimates were made of all species in August.

Fauna were monitored using shelter traps (20 centimetres long x 5 centimetres diameter open-ended plastic cylinders filled with rolled corrugated cardboard) placed in the hedgerow canopy during May for (28 days). Samples were analysed on a preliminary basis into main invertebrate orders.

Differences among treatment means were analysed using ANOVA and differences between ESA and non-ESA hedges were analysed using an independent t-test.

Results

Flora

The mean total species number per treatment plot and the same figure for the permanent quadrats in each plot are presented for 1991 in Table 4.1. It can be seen that most (69 per cent) of species occurring in the plot were recorded in the sample quadrats. There were approximately similar numbers of higher plant species recorded from the ESA and non-ESA hedges (Table 4.2). While all restoration treatments increased plant species diversity, only the two coppiced treatments had significantly ($P<0.05$) more plant species compared to the control (Table 4.1).

Table 4.1 Mean total higher plant species number per treatment plot and per six sample quadrats for each treatment (not including planted species)

Treatment	Control	Lay	Coppice (+ hawthorn)	Coppice (+ mixed species)	Pollard
Treatment means all species/plot	25.7	26.6	31.1	33.6	28.4
Treatment means 6 sample quadrats	17.2	19.3	21.1	22.4	20.6

S.E. treatment means - total species/plot = 1.04 species/6 quadrats = 0.79.

Table 4.2 Mean number of higher plant species per treatment plot per site (1991) in ESA and non-ESA sites

Non-ESA	No. of species	ESA	No. of species
Templepatrick 1	18.8	Carnlough 1	20.0
Templepatrick 2	19.6	Carnlough 2	20.4
Templepatrick 3	20.2	Dromara 1	14.2
Comber	22.6	Dromara 2	18.2
Kilrea	23.0	Hilltown	26.0
Loughgall	22.2	Cushendall	21.8
Omagh	16.2		
Lisbellaw	18.2		
Mean	20.1	Mean	20.1
S.E. mean	1.10	S.E. mean	1.27

Fauna

The mean number of invertebrate orders per site and per treatment plot (Table 4.3A,B) are presented. There was no difference between hedges in the ESA and non-ESA sites. There were significantly more ($P \leq 0.05$) invertebrate orders associated with the laid treatment compared to the control. None of the other treatments differed significantly from the control though there were considerably fewer invertebrate orders found in the control plots compared to the other treatments (Table 4.3B).

**Table 4.3(A) Mean number of invertebrate orders per site (1992) for ESA and
non-ESA sites**

Non-ESA	Orders	ESA	Orders
Templepatrick 1	3.6	Carnlough 1	2.8
Templepatrick 2	3.4	Carnlough 1	1.6
Templepatrick 3	3.6	Dromara 1	3.2
Comber	4.2	Dromara 2	3.6
Kilrea	1.6	Hilltown	4.0
Loughgall	4.6	Cushendall	3.4
Omagh	4.4		
Lisbellaw	4.4		
Mean	3.7	Mean	3.1
S.E. mean	0.324	S.E. mean	0.374

**Table 4.3(B) The effect of restoration treatment on mean number of invertebrate
orders per treatment (1992)**

Treatment	Control	Lay	Coppice (+ hawthorn)	Coppice (+ mixed species)	Pollard
Orders	2.57	4.14	3.64	3.79	3.21
S.E. mean	0.371				

Discussion

The data presented represent the situation in the early stages of a long-term trial and it is
premature to conclude how the different restoration treatments have affected flora and
fauna colonization. However, it is interesting to note that even after one or two years,
definite trends are emerging. The coppiced treatments had a significantly greater
number of plant species compared to the control. This was probably due to more light
entering the hedge-bottom encouraging establishment. On the other hand the laid
hedges in this trial had a significantly more diverse insect fauna compared to the control.
Laying created a denser structure providing a better habitat for a wide range of insect
fauna. As the hedges develop, it is likely that the coppice treatments will 'thicken up'
and become more suitable as a habitat for more insect groups, hence the laid treatment
may only remain more suitable for insects for a relatively short time. Other
characteristics such as diversity of the hedging species, the ground flora, and the
height/thickness ratio may play a greater part in determining the suitability of a hedge as
a habitat for invertebrates. It is likely that, as the invertebrate fauna of a hedge increases
it will be colonized by a wider range of bird species. Conversely, as the coppiced and

pollarded treatments 'thicken up' they will cast more shade and the ground flora associated with the hedge base will become more impoverished (along the lines of the current laid treatment). Overall, even at this early stage it is clear that the flora and fauna of a hedge are highly dynamic and are likely to be greatly influenced by the restoration strategy. However, it is clear that although few treatments were a significant improvement over the control, all treatments gave some improvement even at this stage and it is highly likely that in future, significant improvement will be measured.

Hedges are important as a barrier to stock as well as a wildlife haven and clearly only the laid hedge is likely to achieve this objective in a relatively short time period. However, hedge laying is a time-consuming operation requiring a skill which is not always available.

This trial will enable restoration strategies to be evaluated against each other and considered alongside issues such as skill availability, cost and wildlife enhancement.

Acknowledgements

The authors are grateful to Eugene McBride and the farm staff at Greenmount College for practical assistance and to Roger Martin, Fiona Mulholland, Jenny Thompson, Jill Forsythe, and John Anderson for assistance with sampling and recording.

References

An Foras Forbartha (1985) *The state of the environment.* A report prepared for the Minister for the Environment, Dublin. An Foras Forbartha.

Brooks A. (1988) *Hedging: a practical conservation handbook.* British Trust for Conservation Volunteers, Wallingford.

Carlisle E. (1990) A hedgerow code of practice. *Agriculture in Northern Ireland* **5,** 19-20.

Cooper A, Murray R, McCann T. (1991) Land use and ecological change in Areas of Outstanding Natural Beauty. In: Jeffrey DW, Madden B, eds. *Bioindicators and environmental management.* London: Academic Press, 207-24.

Department of Agriculture for Northern Ireland (1992) *Field boundaries. 1. Field boundaries in the landscape. 2. A hedgerow code of practice. 3. Hedges - planting and aftercare. 4. Managing gappy and overgrown hedges.* Belfast: HMSO.

Department of Agriculture for Northern Ireland (1993) *Statistical review of Northern Ireland agriculture 1992.* Economics and Statistics Division, DANI. Belfast: HMSO.

Department of Agriculture for Northern Ireland/Department of Environment for Northern Ireland (DANI/DOE) (1987) *Hedges on the farm.* Belfast: HMSO.

Graham T. (1979) *Private woodland inventory of Northern Ireland.* Forest Service, Department of Agriculture for Northern Ireland.

Hellewell M. (1991) Managing farm hedges. In: Blyth J, Evans J, Mulch WES, Sidwell C, eds. *Farm woodland management.* Ipswich: Farming Press Books, 45-7.

Mallon E. (1992) Hedges in Northern Ireland. Unpublished dissertation in part fulfilment for degree of BAgr (with Hons), Faculty of Agriculture and Food Science, Queen's University of Belfast.

Rands MRW, Sotherton NW. (1987) The management of field margins for the conservation of game birds. In: Way JM, Greig-Smith PW, eds. *Field margins.* Thornton Heath: British Crop Protection Council, 95-104.

Soper MHR, Carter ES. (1991) *Farming and the countryside.* Ipswich: Farming Press Books.

Webb R. (1985) Farming and the landscape. In: Aalen FHA, ed. *The future of the Irish landscape.* Dublin: Trinity College, 80-92.

CHAPTER 5

FACTORS AFFECTING THE HERBACEOUS FLORA OF HEDGEROWS ON ARABLE FARMS AND ITS VALUE AS WILDLIFE HABITAT

N.D. Boatman[2], K.A. Blake[1], N.J. Aebischer[1] and N.W. Sotherton[1]
[1] *The Game Conservancy Trust, Fordingbridge, Hampshire SP6 1EF, UK*
[2] *The Allerton Research and Educational Trust, Loddington House, Loddington, Leicestershire LE7 9XE, UK*

Introduction

The herbaceous flora of hedgerows

Hedgerows are important as habitat, in their own right and as wildlife refuges; on many farms they form the major non-crop habitat. Most hedgerows contain a strip of herbaceous vegetation in addition to the shrubs of the hedge itself, yet although much has been written about the wildlife value of hedgerows (Pollard *et al.*, 1974; Clements and Tofts, 1992), the emphasis has usually been on the woody component with less consideration of the herbaceous flora. Yet the strip of non-woody plants has certain characteristics not commonly found elsewhere on agricultural land. Firstly, it is often positioned on a bank, providing improved drainage compared to surrounding land. Where soils are not free-draining, a ditch may also be present to aid water removal. Secondly, the hedge provides a combination of shelter and partial shade. Thirdly, hedgerow vegetation on arable farms is not subject to regular defoliation; this normally occurs once a year at most, either in late summer (after harvest) or during winter, i.e. outside the main growing season for most plant species. Similar conditions may be found at some woodland edges, but they are likely to differ in degree. Thus hedgerows have a characteristic community of herbaceous plants which, whilst containing few species which are rare or not found in other habitats (woodland, grassland, etc.) is distinctive in the combination of species, the structure of the vegetation and the prevailing microclimate.

Studies of the herb flora of hedgerows are relatively few. Pollard (1973), Helliwell (1975) and Peterken and Game (1981) have studied the status and potential colonization of woodland herbs in hedgerows. Pollard *et al.* (1974) note that data on comparative frequency of species are not readily available, though they do quote some records from the Floras of Warwickshire (Cadbury *et al.*, 1971) and Wiltshire (Grose, 1957). Boatman and Wilson (1988), reporting a survey of 187 arable field margins in England, noted that the ten most frequently occurring species were common nettle (*Urtica dioica*), hogweed (*Heracleum sphondylium*), creeping thistle (*Cirsium arvense*), barren brome (*Bromus sterilis*), cow parsley (*Anthriscus sylvestris*), cocksfoot (*Dactylis glomerata*), couch (*Elymus repens*), false oat-grass (*Arrhenatherum elatius*), cleavers (*Galium aparine*), and field bindweed (*Convolvulus arvensis*). Cummins *et al.* (1994, this volume) found that the species composition of hedgerow ground flora was mostly related to adjacent land use, rather than to the number of species in the hedge itself or to the hedge management regime.   The most species-rich hedge-bottoms were those next to grasslands, the poorest being those adjacent to arable crops.   A trend towards reduced diversity of herbaceous species was evident between 1978 and 1990, with a shift towards the 'arable' flora.

Herbaceous vegetation as animal habitat

A number of insect groups are known to make use of hedge-bottom vegetation during at least part of their life cycle, and to benefit from a well developed ground flora. Some of these are of economic significance, either as pests, predators of pests or crop pollinators. Traditionally, hedgerows have often been viewed as reservoirs of weeds, pests and diseases, and early studies concentrated on their role in harbouring pest species (Van Emden, 1965).   More recently, greater emphasis has been placed on their value as habitat for beneficial insects.

Pollard (1968*a*) demonstrated that removal of the bottom flora of a hedgerow reduced insect numbers and biomass, particularly predatory species including predacious Heteroptera. A further study (Pollard, 1968*b*) showed that this treatment reduced the numbers of most species of carabid beetles, including species such as *Agonum dorsale* which overwinter in hedge-bottoms and migrate into crops during summer.   The importance of the field boundary flora as overwintering habitat for such polyphagous predators has been demonstrated by Desender (1982), Sotherton (1984, 1985) and Wallin (1985, 1986). Thomas *et al.* (1991, 1992) showed that the density of cocksfoot (*Dactylis glomerata*) tussocks and amount of leaf litter were positively correlated with the distribution of certain predatory beetles.   In fact, for most species important in bio-control, the shrubby component of the hedgerow is unimportant; thus high quality overwintering habitat can be created simply by sowing tussocky grasses on raised banks.

Hoverfly (*Syrphidae*) larvae are voracious aphid predators, but the adults feed on pollen and nectar. Pollen feeding is necessary for females to lay eggs (Schneider, 1969). Van Emden (1965) found that more syrphid eggs were laid near flower-rich edge habitats than in other areas, though Chandler (1968) and Pollard (1971) found that egg distribution was unaffected by proximity of flowers, possibly due to the high mobility of the adult flies.   Cowgill (1989) found that the common species *Episyrphus balteatus* favoured yellow flowers of the Compositae and white Umbelliferae; at the study site both hedgerow perennials and annual weeds were visited, however, on much farmed land weed control practices will severely restrict availability of annual flowers.   Rothery (1992) also found that most hoverflies fed on hedgerow composites and umbellifers, plus the common poppy *Papaver rhoeas*.

Fussell and Corbet (1992) showed the importance of perennial flowering plants to foraging bumble bees (*Bombus* spp.), and recommended the maintenance of undisturbed herbaceous perennial vegetation along field boundaries.   Hedgerows are important butterfly habitats (Thomas, 1986); among species breeding on herbaceous hedgerow plants are the satyrids meadow brown (*Maniola jurtina*), gatekeeper (*Pyronia tithonus*), ringlet (*Aphantopus hyperantus*), and wall brown (*Lasiommata megera*), several species of skipper (Hesperidae), the larval stages of which feed on grasses, the nettle-feeding nymphalids including peacock, (*Inachis io*), small tortoiseshell (*Aglais urticae*) and comma (*Polygonia c-album*), and the pierids orange-tip (*Anthocharis cardamines*) and green-veined white (*Pieris napi*), which feed on crucifers.

Vertebrates which make use of herbaceous hedgerow habitats include small mammals (Tew, 1994, this volume) and adders (Prestt, 1971). The attractiveness of hedgerows to songbirds is primarily determined by the structure of shrub and tree layers (Osborne, 1984; O'Connor, 1984, 1987; O'Connor and Shrubb, 1986). However, Osborne (1982) found that the number of bird species present was related to species' richness of the herb layer, and Arnold (1983) also found that the number of herbaceous species present was one of the five dominant variables influencing bird numbers in summer. Passerines which nest on the ground or in low vegetation in hedges include whitethroat (*Sylvia communis*), yellowhammer (*Emberiza citrinella*) and reed bunting (*E. shoeniclus*).

Hedgerows are the major nesting habitat for both grey and red-legged partridges (*Perdix perdix* and *Alectoris rufa*) in Britain (Rands, 1986, 1987), and studies have shown that breeding density of the grey partridge is related to the amount of dead grass in the base of the hedge, whilst for red-legged partridges the amount of nettle was important (Rands, 1986). Both grey and red-legged partridges chose to nest where amounts of dead grass, bramble and leaf litter were greater; grey partridges preferred hedgerows where a high bank was present, whilst red-legged partridges favoured sites with a higher proportion of nettle than adjacent areas (Rands, 1988). Ground vegetation characteristics also influenced the degree of nest predation (Rands, 1988). Rands made no attempt in these studies to distinguish plant species other than as the categories of 'grass', nettle, bramble, and 'ground cover'.

Management of hedge-bottom vegetation

Hedge-bottoms or banks may be subject to a variety of forms of management, depending on the form of adjacent land use, the traditions of the area and the perceptions of the individual farmer. The most benign is probably an occasional trim with a flail mower at the same time as the hedgerow shrubs are cut; neglected hedgerows may not even receive this. At the other end of the spectrum is the deliberate annual application of broad-spectrum herbicides in an attempt to control weeds and maximize cropping area. Surveys of farmers at agricultural shows have shown that a large proportion (around 60 per cent) of farmers use herbicides in their field boundaries (Marshall and Smith, 1987; Boatman, 1989). In between these two extremes are various degrees of disturbance, either as active management (e.g. frequent cutting) or as a by-product of farming operations such as cultivations, misplaced fertilizer or pesticide drift. The effects of most of these management influences have been little studied, though a number of recent papers have addressed the implications of herbicide drift (Marrs *et al.*, 1989, 1991a,b; Breeze *et al.*, 1992). Parr and Way (1988) showed that finer grasses and smaller herbs increased at the expense of larger or coarser species with increased cutting frequency on road verges, but the cuts were carried out during the summer months when most arable field margins are inaccessible due to the presence of crops in the field and are, therefore, not cut. Watt *et al.* (1990) showed that cutting in June reduced seed return of some annuals, though not barren brome (*Bromus sterilis*); again cutting at this time of year would only be practicable on wide margins, and is not currently common practice.

Disturbance of field margin vegetation is generally perceived to result in a shift from a benign perennial flora, to increased dominance by annual weedy species which may invade the crop. However, the effects of different degrees and types of disturbance have been little studied. There is considerable scope for further work on the effects of management and mismanagement on the herbaceous component of hedgerows.

Current studies

The work described here falls into two parts. Firstly, survey data are used to characterize the ground flora of arable field boundaries in southern England and the effect of presence or absence of a hedge. Secondly, the importance of plant species composition in determining habitat preference is investigated with respect to nest site choice in the grey partridge.

Materials and methods

(a) *Botanical surveys*

Surveys of arable field boundaries were carried out on farms belonging to Cereals and Gamebirds Project subscribers across the south of England. Nine farms were surveyed in 1987 and 21 in 1988. Results from 103 boundaries from all nine farms surveyed in 1987, and 93 boundaries from ten of the farms surveyed in 1988, are reported here.

The survey included the cropped headland as well as the field boundary vegetation, but only results from field boundaries are reported here. All cereal field margins with conservation headlands were surveyed on each farm and an equivalent number of margins with sprayed headlands were also surveyed. Sprayed and conservation headlands were paired with respect to aspect and soil type as far as possible.

Ten 0.25 square metre quadrats were placed centrally in the field boundary strip, 5 metres (1987) or 10 metres (1988) apart in the selected boundaries. The section to be recorded was chosen so that gateways, individual overhanging trees or other features likely to influence vegetation locally were avoided.

The presence or absence of each species in the ten quadrats was examined in relation to field boundary type (hedge >2 metres, hedge <2 metres, shelter belt, woodland edge, fence, grass strip etc.), and width of herbaceous vegetation. For each boundary, the presence/absence data were expressed as the proportion of quadrats in which a species was present. After angular transformation, they were used as the dependent variable in a stepwise linear multiple regression against the boundary variables (boundary type, width), to determine the effect of these factors (if any) on frequency of the individual species.

(b) *Effect of plant species composition on choice of nest site by grey partridge* (Perdix perdix)

In 1991, two radio-tracking studies of hen grey partridges were carried out between February and July, with the primary aim of studying causes of mortality during the nesting season. (Reynolds *et al.*, 1992) They took place on the Manydown Estate, near Basingstoke and on Salisbury Plain. A unique feature of these studies was that partridges were radio-tagged before nesting occurred, thus providing for the first time an unbiased sample of the range of habitats chosen for nest sites. Previous studies have relied on finding nests by searching hedgerows (Rands, 1986, 1987, 1988). The botanical composition of the nest site area was recorded, as well as physical structure of the vegetation and a number of other habitat variables (distance to hedge, distance to nearest tree, distance to crop edge, adjacent land use, ground vegetation height, and for nests in field boundaries, boundary type, aspect, height, width, slope of bank, bank height, nest height, nearest gap in hedge, and shrub species in hedge).

At each study site, an equivalent number of 'non-nest' sites were recorded for comparative purposes, at two levels of resolution on a landscape scale. 'Low resolution' sites were selected at random within the study area from all habitat areas of the same type as that chosen for nesting. For this purpose, habitats were grouped into field boundary, crop and 'other' (rough grass, scrub, nettles, game cover). 'High resolution' sites were selected from the same area (hedgerow, field, patch of rough grass etc.) as the paired nest site.

Plant species were recorded at nest sites as soon as the nest was vacated, either due to hatching or predation. Four 0.25 square metre quadrats were recorded around each nest, leaving the central 0.25 square metres unrecorded as the vegetation was disturbed by the presence of the nest. Percentage cover of each species present was estimated by eye in each quadrat, viewed from above. Non-nest sites were recorded in a similar manner.

Data were analysed by logistic stepwise multiple regression, comparing nests and non-nests. Initial analysis incorporating all habitat types showed no significant differences, so data from cropped sites were excluded and remaining data re-analysed. The results of this analysis are presented here.

For the purposes of analysis, sites with nests were coded 1, those without were coded 0. Factors related to nest presence/absence were identified by logistic stepwise multiple regression (Nicholls, 1989), using the statistical package GENSTAT 5 (Genstat 5 Committee, 1987). The analysis reported here was restricted to field boundary and similar habitats, and incorporated a 'site' factor at step 0 to take into account possible differences between Manydown and Salisbury Plain.

Results

(a) *Botanical surveys*

The frequency of occurrence in terms of number of field boundaries and mean number of quadrats per boundary for the most common species is shown in Table 5.1. Six species were recorded in more than 50 per cent of boundaries, and three species occurred in more than 30 per cent of quadrats. The ratio of the mean number of quadrats to the number of boundaries gives a crude measure of the type of distribution of species; those occurring in a high number of quadrats in relation to number of boundaries would be expected to tend towards a uniform distribution within boundaries (e.g. *Arrhenatherum elatius, Poa trivialis, Urtica dioica, Agrostis stolonifera, Elymus repens, Bromus sterilis*), whilst those giving a low ratio would tend to have a clumped distribution (e.g. *Lamium album, Cirsium arvense, Lolium perenne*).

Table 5.1   Frequency of occurrence of species occurring in 20 per cent or more of field boundaries in surveys of herbaceous vegetation

	% Field boundaries	Mean no. quadrats per field boundary (max = 10)	% Quadrats / % boundaries
Urtica dioica	74	3.35	0.46
Poa trivialis	69	3.42	0.49
Arrhenatherum elatius	67	3.72	0.55
Bromus sterilis	62	2.69	0.44
Galium aparine	60	2.33	0.39
Cirsium arvense	56	1.55	0.28
Dactylis glomerata	47	1.45	0.31
Elymus repens	45	2.06	0.45
Heracleum sphondylium	40	1.35	0.33
Convolvulus arvensis	38	1.25	0.33
Anthriscus sylvestris	35	1.16	0.33
Glechoma hederacea	29	0.93	0.32
Lolium perenne	26	0.73	0.29
Lamium album	22	0.46	0.21
Agrostis stolonifera	20	0.89	0.45

The majority of these species are polycarpic perennials, with two exceptions, *Galium aparine* and *B. sterilis*. Most share certain common characteristics; they are competitive (C, CR or CSR strategists; Grime, 1974) and successful colonizers, characteristic of high soil fertility and often associated with disturbance. They are tolerant of moderate shade, but not of defoliation.

Species whose occurrence was significantly affected by hedge type are shown in Table 5.2. Only *A. stolonifera* was significantly more frequent where no hedge was present. *Anthriscus sylvestris, Galium aparine, Hedera helix, Heracleum sphondylium,* and *Urtica dioica* were more common where hedges were present; *Glechoma hederacea* was most common next to tall hedges (higher than two metres).

Table 5.2   Effect of presence and height of hedge on herbaceous plant species

Data are mean number of quadrats per field boundary (maximum=10); data in parentheses are angular transformed (10 x values with standard errors)

Species	No hedge	Short hedge	Tall hedge
Anthriscus sylvestris	0.80 (9.25+2.73)	1.20 (12.83+2.39)	1.67 (16.41+2.16)
Galium aparine	1.33 (15.13+3.47)	3.25 (29.71+3.03)	2.42 (24.00+2.74)
Glechoma hederacea	0.56 (6.13+2.45)	0.42 (4.86+2.14)	1.04 (11.29+1.93)
Hedera helix	0.13 (1.13+1.66)	0.34 (4.53+1.45)	0.44 (5.02+1.31)
Heracleum sphondylium	0.47 (5.83+3.11)	2.03 (19.83+2.72)	1.26 (13.99+2.46)
Urtica dioica	1.67 (17.22+3.81)	3.07 (30.45+3.33)	4.51 (41.28+3.01)

Several species showed a positive relationship with verge width (Table 5.3); these were all grassland species, intolerant of shading, but (with the exception of *Holcus lanatus*), tolerant of regular defoliation. Three species, *B. sterilis*, *G. aparine* and *E. repens*, were negatively related to verge width. *G. aparine* is commonly found germinating in the bare soil beneath hedges, which it is then able to scramble up or through.   Where verges are narrow, disturbance from agricultural operations may encourage its spread across the verge from the hedge itself. All three species thrive in undisturbed habitats not subject to regular defoliation, and are potentially serious arable weeds.

Table 5.3   Regression parameters for species showing significant relationship between frequency of occurrence and verge width (analysis carried out on angular transformed data)

Species	Regression coefficient	Constant	$F_{1,193}$	P
Agrostis stolonifera	2.06	4.07	29.1	<0.001
Festuca rubra	1.51	2.77	23.8	<0.001
Holcus lanatus	1.05	3.03	15.5	<0.001
Lolium perenne	1.88	4.00	32.6	<0.001
Ranunculus repens	0.60	2.94	8.0	<0.01
Bromus sterilis	-1.51	29.40	7.1	<0.01
Elymus repens	-1.22	22.53	5.0	<0.05
Galium aparine	-1.44	26.17	8.7	<0.01

A simple constellation diagram based on correlation coefficients gives some further insight into relationships between the species themselves (Figure 5.1). The diagram is dominated by a central cluster of dicotyledonous species, notably umbellifers and labiates plus *G. aparine* and the grasses *E. repens* and *B. sterilis*. Peripheral to this are coarse perennial grasses characteristic of tall herb communities: *A. elatius, Dactylis glomerata* and *H. lanatus*, together with *Festuca rubra* which forms a non-flowering understorey in such communities (Grubb, 1982). A further group, consisting of three species characteristic of grazed grassland, has some positive correlations with *D. glomerata* and *H. lanatus*, but there are a number of negative correlations between these three species and species present in the central cluster. These three grassland species *A. stolonifera*, *L. perenne* and *Ranunculus repens*, were all positively related to verge width.

Figure 5.1   Constellation diagram showing correlations between species from botanical survey

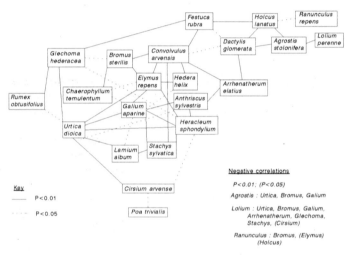

(b) *Choice of nest site by grey partridge*

A wide range of cover types was selected, with grass banks, rough grass and hedges as the main choices (Table 5.4). A number of species were selected during the 'low resolution' regression analysis, in addition to dead grass and leaf litter (Table 5.5). The amounts of dead grass and leaf litter were also found to be important factors influencing nest site selection by Rands (1988). The species which were positively associated with nest sites as opposed to 'non-nests' were generally characteristic of mixed, tall herb communities subject to some degree of disturbance. It may be that such occasional disturbance creates gaps in the vegetation, which are suitable for positioning a nest. Of the two species negatively associated with nest sites *Trifolium repens* is characteristic of short, regularly defoliated grassland, is intolerant of shade, and rapidly suppressed by tall vegetation; *B. sterilis* is also unable to persist in tall herb communities and is characteristic of a high degree of disturbance. It is often abundant in a strip of partially cultivated ground between perennial hedge-bottom vegetation and a crop.

Table 5.4   Habitat types selected for nesting by radio-tagged partridges

Habitat type	Nests	%
Hedge	4	16
Grass bank	7	28
Shelter belt	2	8
Rough grass	5	20
Scrub	3	12
Nettles	2	8
Game cover[a]	2	8
Total	25	

[a] One in Jerusalem artichokes, one in canary grass.

Table 5.5   Logistic stepwise multiple regression analysis of nest versus 'non-nest' sites for grey partridge

Variables are listed in order of selection, with the corresponding residual deviances and degrees of freedom (approximate chi-squared goodness-of-fit). The regression coefficients and tests of significance are given for the final equation.

Variable	Step	Deviance	d.f.	Regression coefficient[a]	X^2	P
(a)Low Resolution						
Site	0	114.49	103	-0.252	0.10	NS
Dead grass	1	104.08	102	0.125	17.32	<0.001
Trifolium repens	2	94.07	101	-0.342	5.62	<0.05
Rumex crispus	3	86.17	100	0.859	9.77	<0.01
Galium aparine	4	79.12	99	0.087	7.23	<0.01
Stellaria holostea	5	73.54	98	1.310	10.16	<0.01
Leaf litter	6	67.95	97	0.210	7.33	<0.01
Lamium album	7	63.10	96	0.253	6.49	<0.05
Hypericum perforatum	8	58.63	95	0.403	4.35	<0.05
Malva sylvestris	9	54.11	94	3.100	6.56	<0.05
Bromus sterilis	10	49.66	93	-0.220	4.45	<0.05
(b) High resolution						
Site	0	69.31	48	0.105	0.03	NS
	1	61.44	47	-2.360	13.87	<0.001
Arctium minus	2	59.06	46	2.430	7.05	<0.01
Rumex crispus	3	51.24	45	1.220	4.81	<0.05

[a] A positive coefficient indicates a greater association with nests than non-nests, and vice-versa.

The 'high resolution' analysis revealed only a negative association of nest sites with *T. repens* and a positive association with *Arctium minus* and *R. crispus*, both of which are characteristic of moderate levels of disturbance (Table 5.5).

Discussion

Hedgerows next to arable fields typically have a ground flora which is impoverished to a greater or lesser extent largely due to the impact of agricultural operations in the adjacent field, and is often dominated by a few species of coarse herbs such as *Urtica dioica*, *Arrhenatherum elatius* and *Heracleum sphondylium*. Co-dominant with such species are others which have the potential to become arable weeds, e.g. the perennials *Poa trivialis*, *Cirsium arvense*, *Elymus repens*, *Convolvulus arvensis*, and the annuals *Bromus sterilis* and *Galium aparine*. Even *A. elatius* has a weedy form, known as onion couch though this has a restricted distribution and factors governing its occurrence have not been properly elucidated (Cussans *et al.*, 1992). Farmers, therefore, often view hedgerows as a reservoir of weeds, and treat them accordingly, though they may unwittingly encourage the very species they are attempting to control. Conservationists, particularly botanists, often view the hedgerow ground flora with a similar jaundiced eye; the species present are common, widespread and not 'interesting'. Consequently, recent research has often been directed at attempts to replace existing vegetation with a species mixture considered more 'desirable', rather than at the management of the indigenous flora (Smith and MacDonald, 1989; Marshall and Nowakowski, 1991).

Such an approach may provide benefits in terms of weed control, if only by widening the verge and providing an incentive to more careful management. The association between species characteristic of mown or grazed grassland and verge width in the current study probably indicates the presence of wide grass verges which are kept regularly mown, a policy deliberately adopted by a few arable farmers. The negative association between *B.sterilis*, *G. aparine*, *E.repens* and verge width indicates that these species at least are kept in check by this practice. An alternative approach, controlling weed species by the use of selective herbicides which do not damage non-target species, gives good results in the short-term, but its effectiveness in the longer term (over periods of two years or longer) has yet to be fully tested (Boatman, 1989, 1992). There may be some scope for complementary use of these two approaches (Marshall and Nowakowski, 1991).

Whether sowing introduced species provides animal habitats of equivalent value to that provided by the indigenous vegetation depends on how much attention is given to the requirements of the animal species or groups involved. Structure may be as important or more important than actual botanical composition, as suggested by the wide variety of types of cover chosen by nesting partridges. Vegetation height is clearly important, and the presence of dead grass and leaf litter have previously been established as beneficial (Rands, 1988). The only obvious connections between the plant species associated more frequently with nests than non-nests were that all are broad-leaved species and most are characteristic of some degree of disturbance. Thus, a continuous grass sward, even of tall tussocky species, may be less suitable than one subject to intermittent disturbance, allowing the establishment of species such as *Rumex crispus*,

Arctium minus and *Galium aparine*. The presence of the shrub component of the hedgerow does not seem to be necessary; more partridges nested in grassy banks or rough grass than next to hedges or in scrub.

Several species were more frequent where hedges were present, in comparison with boundaries containing only herbaceous vegetation. *Glechoma hederacea* and *Hedera helix* characteristically occur in shaded or partially shaded habitats, and do not persist in unshaded tall herb communities. *Galium aparine*, as stated earlier, benefits from bare soil beneath hedges where it can germinate and establish free from competition, and from the hedge itself as a structural support.

The reasons why *Anthriscus sylvestris*, *Heracleum sphondylium* and *Urtica dioica* tend to be more frequent where hedges are present are not so clear. All are capable of persisting in unshaded tall herb communities, but may benefit from partial shade in competition with grasses. They are intolerant of regular defoliation, and verges with no shrubby cover may be defoliated more frequently in the cause of 'tidiness'.

The method of analysis used is subject to certain limitations which are overcome by recent multi-variate techniques. Regression analysis assumes linear relationships and considers only one species at a time. Ordination techniques allow analysis of relationships between species within plant communities, but used alone do not provide for statistical tests of relationships between species and environmental variables.

However, the advantages of regression and ordination analysis are combined in canonical ordination techniques, e.g. canonical correspondence analysis, (ter Braak, 1988) an extension of detrended correspondence analysis (DECORANA) (Hill, 1979). Further analysis of the data set using such techniques could enable a wider interpretation of between-species relationships and community responses to environmental parameters to be made.

In conclusion, the herbaceous vegetation of hedgerows, whilst often degraded in terms of botanical diversity, provides a valuable resource in terms of wildlife habitat. A greater understanding of the factors affecting botanical composition, combined with research into the habitat requirements of beneficial or otherwise desirable animal taxa, as with partridges and predatory beetles, should enable the development of management techniques to utilize this potential more effectively whilst minimizing any adverse impact on agriculture.

References

Arnold GW. (1983) The influence of ditch and hedgerow structure, length of hedgerows, and area of woodland and garden on bird numbers on farmland. *Journal of Applied Ecology* **3**, 731-50.

Boatman ND. (1989) Selective weed control in field margins. *1989 Brighton Crop Protection Conference - Weeds* **2**, 785-95.

Boatman ND. (1992) Improvement of field margin habtitat by selective control of annual weeds. *Aspects of Applied Biology* **29**, *Vegetation management in forestry, amenity and conservation areas*, 431-36.

Boatman ND, Wilson PJ. (1988) Field margin management for game and wildlife conservation. *Aspects of Applied Biology* **16**, *The practice of weed control and vegetation management in forestry, amenity and conservation areas*, 53-61.

Breeze V, Thomas G, Butler R. (1992) Use of a model and toxicity data to predict the risks to some wild plant species from drift of four herbicides. *Annals of Applied Biology* **121**, 669-77.

Cadbury DA, Hawkes JG, Readett RC. (1971) *A computer mapped flora: a study of the county of Warwickshire*, London: Academic Press.

Chandler AEF. (1968) Some factors influencing the site and occurrence of oviposition by aphidophagous Syrphidae (Diptera). *Annals of Applied Biology* **61**, 435-46.

Clements DK, Tofts RJ. (1992) *Hedgerow evaluation and grading system (HEGS). A methodology for the ecological survey, evaluation and grading of hedgerows. Test Draft, September 1992*, Cirencester: Countryside Planning and Management. 61pp.

Cowgill S. (1989) The role of non-crop habitats on hoverfly (Diptera: Syrphidae) foraging on arable land. *1989 Brighton Crop Protection Conference - Weeds* **3**, 1130-38.

Cummins RP, French DD. (1994) Floristic diversity, management and associated land use in British hedgerows. In: Watt TA, Buckley GP, eds. *Hedgerow management and nature conservation.* Wye College Press, Wye College, University of London, 95-106.

Cussans JW, Miller ACE, Morton AJ, Cussans GW. (1992) *Arrhenatherum elatius* - a potential problem under set-aside? In: Clarke J, ed. *Set-aside.* BCPC Monograph No. **50**. Farnham: British Crop Protection Council, 147-50.

Desender K. (1982) Ecological and faunal studies on Coleoptera in agricultural land. II. Hibernation of Carabidae in agro-ecosystems. *Pedobiologia* **23**, 295-303.

van Emden HF. (1965) The effect of uncultivated land on the distribution of cabbage aphid (*Brevicoryne brassicae*) on an adjacent crop. *Journal of Applied Ecology* **2**, 171-96.

Fussell M, Corbet SA. (1992) Flower usage by bumble-bees: a basis for forage plant management. *Journal of Applied Ecology* **29**, 451-65.

Genstat 5 Committee (1987) *Genstat 5 Reference Manual*. Oxford: Clarendon Press.

Grime JP. (1974) Vegetation classification by reference to strategies. *Nature* **250**, 26-31.

Grose D. (1957) *The Flora of Wiltshire*. Devizes, Wiltshire: Archaeological and Natural History Society.

Grubb PJ. (1982) Control of relative abundance in roadside *Arrhenatherum*: results of a long-term garden experiment. *Journal of Ecology* **70**, 845-65.

Helliwell DR. (1975) The distribution of woodland plant species in some Shropshire woods. *Biological Conservation* **7**, 61-72.

Hill MO. (1979) *DECORANA: a FORTRAN program for detrended correspondence analysis and reciprocal averaging.* Section of Ecology and Systematics, Cornell University, Ithaca, New York.

Marrs RH, Williams CT, Frost AJ, Plant RA. (1989) Assesment of the effects of herbicide drift on a range of plant species of conservation interest. *Environmental Pollution* **59,** 71-86.

Marrs RH, Frost AJ, Plant RA. (1991*a*) Effects of herbicide spray drift on selected species of nature conservation interest: the effects of plant age and surrounding vegetation structure. *Environmental Pollution* **69,** 223-35.

Marrs RH, Frost AJ, Plant RA. (1991*b*) Effect of mecoprop drift on some plant species of conservation interest when grown in standardized mixtures in microcosms. *Environmental Pollution* **73,** 25-42.

Marshall EJP, Nowakowski M. (1991) The use of herbicides in the creation of a herb-rich field margin. *1991 Brighton Crop Protection Conference - Weeds,* 655-60.

Marshall EJP, Smith BD. (1987) Field margin flora: interactions with agriculture. In: Way JM, Greig-Smith PW, eds. *Field margins.* BCPC Monograph No. **35.** Thornton Heath: British Crop Protection Council, 23-33.

Nicholls AO. (1989) How to make biological surveys go further with generalized linear models. *Biological Conservation* **50,** 51-75.

O'Connor RJ. (1984) The importance of hedges to songbirds. In: Jenkins D, ed. *Agriculture and the environment.* ITE Symposium No. **13.** Cambridge: Institute of Terrestrial Ecology.

O'Connor RJ. (1987) Environmental interests of field margins for birds. In: Way JM, Greig-Smith PW, eds. *Field margins.* BCPC Monograph No. **35.** Thornton Heath: British Crop Protection Council, 35-48.

O'Connor RJ, Shrubb M. (1986) *Farming and birds.* Cambridge: Cambridge University Press.

Osborne PJ. (1982) *The effects of Dutch Elm disease on farmland bird populations.* DPhil Thesis, Oxford University.

Osborne PJ. (1984) Bird numbers and habitat characteristics in farmland hedgerows. *Journal of Applied Ecology* **21,** 63-82.

Parr TW, Way JM. (1988) Management of roadside vegetation: the long-term effects of cutting. *Journal of Applied Ecology* **25,** 1073-87.

Peterken GF, Game M. (1981) Historical factors affecting the distribution of *Mercurialis perennis* in central Lincolnshire. *Journal of Ecology* **69,** 781-96.

Pollard E. (1968*a*) Hedges. II. The effect of removal of the bottom flora of a hawthorn hedgerow on the fauna of the hawthorn. *Journal of Applied Ecology* **5,** 109-23.

Pollard E. (1968*b*) Hedges. III. The effect of removal of the bottom flora of a hawthorn hedgerow on the Carabidae of the hedge bottom. *Journal of Applied Ecology* **5,** 125-39.

Pollard E. (1971) Hedges. IV. Habitat diversity and crop pests: a study of *Brevicoryne brassicae* and its syrphid predators. *Journal of Applied Ecology* **8,** 751-80.

Pollard E. (1973) Hedges. VII. Woodland relic hedges in Huntingdon and Peterborough. *Journal of Ecology* **61,** 343-52

Pollard E, Hooper MD, Moore NW. (1974) *Hedges.* London: Collins.

Prestt I. (1971) An ecological study of the viper *Vipera berus* in southern Britain. *Journal of Zoology* **164**, 373-418.

Rands MRW. (1986) Effect of hedgerow characteristics on partridge breeding densities. *Journal of Applied Ecology* **23**, 479-87.

Rands MRW. (1987) Hedgerow management for the conservation of partridges *Perdix perdix* and *Alectoris rufa. Biological Conservation* **40**, 127-39.

Rands MRW. (1988) The effect of nest site selection on nest predation in grey partridge *Perdix perdix* and red-legged partridge *Alectoris rufa. Ornis Scandinavica* **19**, 35-40.

Reynolds JC, Dowell SDD, Brockless MH, Blake K, Boatman ND. (1992) Tracking partridge predation. *The Game Conservancy Review of 1991* **23**, 60-2.

Rothery F. (1992) Hoverfly foraging on hedgerow flowers. In: *Proceedings Institute of Biological Control Conference, Rennes. IOBC/WPRS Bulletin No. 5.*

Schneider F. (1969) Bionomics and physiology of aphidophagous Syrphidae. *Annual Review of Entomology* **14**, 103-24.

Smith H, Macdonald DW. (1989) Secondary succession on extended arable field margins: its manipulation for wildlife benefit and weed control. *1989 Brighton Crop Protection Conference - Weeds,* 1063-68.

Sotherton NW. (1984) The distribution and abundance of predatory arthropods overwintering on farmland. *Annals of Applied Biology* **105**, 423-9.

Sotherton NW. (1985) The distribution and abundance of predatory Coleoptera overwintering in field boundaries. *Annals of Applied Biology* **106**, 17-21.

ter Braak CJF. (1988) CANOCO - an extension of DECORANA to analyse species-environment relationships. *Vegetatio* **75**, 159-60.

Tew TE. (1994) Farmland hedgerows: habitat, corridors or irrelevant? A small mammal's prespective. In: Watt TA, Buckley GP, eds. *Hedgerow management and nature conservation.* Wye College Press, Wye College, University of London, 80-94.

Thomas JA. (1986) *RSNC guide to butterflies of the British Isles*, Twickenham, Middlesex: Country Life Books.

Thomas MB, Wratten SD, Sotherton NW. (1991) Creation of 'island' habitats in farmland to manipulate populations of beneficial arthropods: predator densities and emigration. *Journal of Applied Ecology* **28**, 906-17.

Thomas MB, Wratten SD, Sotherton NW. (1992) Creation of 'island' habitats in farmland to manipulate populations of beneficial arthropods: predator densities and species composition. *Journal of Applied Ecology* **29**, 524-31.

Wallin H. (1985) Spatial and temporal distribution of some abundant carabid beetles (Coleoptera: Carabidae) in cereal fields and adjacent habitats. *Pedobiologia* **28**, 19-34.

Wallin H. (1986) Habitat choice of some field inhabiting carabid beetles (Coleoptera: Carabidae) studied by recapture of marked individuals. *Ecological Entomology* **11**, 457-66.

Watt TA, Smith H, Macdonald DW. (1990) The control of annual grass weeds in fallowed field margins managed to encourage wildlife. In: *Proceedings European Weed Research Society Symposium on Integrated Weed Management in Cereals, Helsinki,* 187-96.

CHAPTER 6

CONTROL OF BIODIVERSITY IN HEDGEROW NETWORK LANDSCAPES IN WESTERN FRANCE

F. Burel[1] and J. Baudry[2]
[1] *Laboratoire d'Evolution des Systèmes Naturels et Modifiés, Université de Rennes 1, Campus de Beaulieu, 35042 Rennes Cedex, France*
[2] *Institut National de la Recherche Agronomique, Département de recherche sur les systèmes agraires et le développement, 14170, Lieury, France*

Introduction

The landscapes of western France are characterized by hedgerows that may be dated as far back as prehistoric periods (Giot *et al.*, 1979). Landscape history has been a succession of establishment and removal periods (Meyer, 1972). From 1840 to 1914, the cultivation of common moorland led to the main establishment process, when the network of hedgerows was at its highest density. More recently, as in many other regions in western Europe (Leonard and Cobham, 1977; Agger and Brandt, 1988), agricultural evolution led to field enlargement and, consequently, to hedgerow removal, either by individual initiatives, or within land consolidation programmes.

The ecological consequences of these drastic and rapid landscape changes were an increase in flooding, soil erosion and wind damage and the scientific community has, therefore, been stimulated by a public and governmental concern to initiate an extensive research programme on the ecology, economy and history of these vanishing landscapes (INRA *et al.*, 1976). These ecological studies, which focused mainly on the hedgerow, and hedgerow-field levels, underlined the role of hedgerows in controlling the biological diversity of such landscapes (Lefeuvre *et al.*, 1976). The results are broadly similar to researches taking place at the same time in the UK (Pollard *et al.*, 1974). Within hedges, the faunal diversity is high due to micro-habitat heterogeneity (trunks, stones, ditches), the complexity of the vegetation vertical structure (number of layers, plant species present, pruning regimes), and to host plant diversity. The fauna is not specific, but consists of a mixture of species derived from different environments, particularly woods. For the different species that spend part or all of their life cycle in hedgerows, the hedgerow may be considered as (a) the habitat for those that are restricted to it, (b) a temporary refuge for those that feed or spend part of their life in adjacent fields, (c) a complementary feeding area, or (d) a reservoir for propagules. Hedgerows influence the spatial distribution of insects in adjacent fields, both because they are sources of species and because they influence field microclimate (Guyot and Verbrugghe, 1976).

Such studies have continued in the 1980s, with the result that scientific knowledge is now sufficient to manage individual hedgerows either to maximize the diversity of one group of species (Pollard, 1968; Lewis, 1969; Arnold, 1983; Yahner, 1983; Osborne, 1984; Soltner, 1978; Lack, 1988) or as part of an integrated pest control programme.

However, this research has neglected hedgerows as elements of whole landscapes and has not considered interactions among hedgerows or between hedgerows and other uncultivated elements as possible factors controlling biodiversity.

Hedgerows and landscape

The existence of interactions between landscape elements is a central paradigm of landscape ecology (Forman, 1983). Within the framework of this new field (Tjallingii and de Veer, 1981; Brandt and Agger, 1984; Schreiber, 1988), research has focused on the impact of landscape structure and dynamics on ecological processes such as fluxes of matter, nutrients and species. Effects of habitat fragmentation and landscape heterogeneity have been integrated into the subject of population biology (Opdam, 1991; Merriam, 1990) using the concept of the metapopulation to describe population dynamics in fragmented habitats (Levins, 1970; Hanski, 1989). Linear landscape elements are corridors enhancing movements (Saunders and Hobbs, 1991), barriers (Mader, 1988) or ecotones (Naiman and Décamps, 1990). In hedgerow network landscapes research has been initiated on the role of spatial structure (heterogeneity, connections, length of hedges or distance from forest source) on species distribution (Forman and Baudry, 1984). It has been demonstrated for several taxonomic groups such as plants (Baudry, 1985), carabids (Burel, 1989, 1992) and birds (Balent and Courtiade, 1992) that the actual biological diversity of one hedgerow is the result of constraints exerted at the landscape level, as well as of its own characteristics.

In this paper we illustrate, by a review of already published research, how both connection, and distance from population sources influence populations of different taxa. We discuss how the role of hedgerows as corridors is dependent upon their structure and on the component species. Hedgerow network landscapes may be viewed as woody fragmented habitats (Burgess and Sharpe, 1981) and the metapopulation concept may be applied to some species. We concentrate on forest species (i.e. species that are significantly more frequent in forests than in other habitats) as they are the most threatened by hedgerow clearance in landscapes such as ours, in Brittany France, where woodland is scarce.

In the following sections the landscape scale is emphasized, allowing us to present, in our conclusion, some general principles of landscape design and management which can be used to preserve hedgerow species diversity.

It should be noted that in landscape ecology, experimentation is very difficult because phenomena occur at coarse spatial and time-scales. It is almost impossible, therefore, to use parametric statistics, as stressed by many authors because, for example, samples are not independent and data are non-normal. Landscapes are extremely complex systems with many (non-linear) correlations among variables.

Are hedgerows good corridors for forest species?

As shown by earlier studies, many forest species may be found in hedgerows, including carabid beetles (Thiele, 1977), birds (Wegner and Merriam, 1979), butterflies (Pollard *et al.*, 1974), snails (Cameron *et al.*, 1980), and shrubs (Helliwell, 1975). How are these populations related to each other in hedgerow networks, and to the main source of propagules, the forests? Are hedgerows good corridors (Forman and Godron, 1986)? Corridors may fulfil different functions in the landscape (Bennett, 1990): they may be a habitat, they may ease the movement of plants and animals across hostile environments or they may be a source of biotic or abiotic effects on the surrounding elements.

In Brittany, north of Rennes, we studied an agricultural area adjacent to a large (7,000 hectare) ancient forest to test the efficiency of hedgerows as corridors for forest species. We chose four groups that differ in their way of dispersal, their mobility, and the scale at which they perceive landscape spatial structure: plants, spiders, carabids, and birds (Burel, 1988, 1989, 1991).

The forest is composed of two main stands: the first one consists of dense hornbeam (*Carpinus betulus*) coppice, while in the second oak and beech trees (*Quercus robur* and *Fagus sylvatica*) are grown for timber. In the rural area that extends 1.5 kilometres from the forest edge and is 1 kilometre wide, the fields are mostly meadows and pastures. Trees in hedgerows separating fields are mainly pruned oaks, growing on the top of earthen banks. The forest, all hedgerows and some fields were sampled for all species groups.

Among these four different groups the method of dispersal appears to determine their reaction to landscape structure (Burel, 1991).

Corridor efficiency and species group

In the study area, 72 taxa of spiders have been identified. Their spatial distribution reflects the hedgerow structure, but there is no significant relationship with distance from the forest. This is because the 'propagule' rain of web-weaving spiders, being mainly wind dispersed (Southwood, 1962), reaches every landscape element whatever the landscape structure. Colonization occurs when local conditions are favourable.

In contrast, some herbaceous woodland plant species (Helliwell, 1975; Peterken and Game, 1981) and forest carabid beetles (Den Boer, 1981) have a much lower dispersal ability. Their presence in hedgerows depends on the proximity of a source. Our results show that landscape structure (distance from the forest, connectedness among favourable elements) controls the spatial distribution for these two groups of species (Burel, 1991). Structural connections among hedgerows (connection) permit functions such as species movement from one element to another (connectivity) (Baudry and Merriam, 1988).

Only some carabid species inhabiting the forest are found in hedgerows (Burel, 1989). Some are restricted to the forest like *Abax parallelus*, while some are only found in hedgerows less than 1 kilometre from the forest like *Abax ovalis*, their abundance decreasing gradually as if hedgerows were acting as forest peninsulas (Milne and Forman, 1986). The absence of these species within the agricultural landscape may simply be due to their inability to move far from the forest whatever type of hedgerow management is carried out. For the last group of species (e.g. *Abax ater*) hedgerows are good corridors and were found at the limit of the study area.

Among the 39 forest plant species, only three (*Milium effusum*, *Melampyrum pratense* and *Viburnum opulus*) were not present in hedgerows and there was no sign of a progressive colonization from the forest. One may hypothesize that there has been sufficient time for species to colonize any favourable habitat, but on the contrary, studies of shrub distribution in a landscape composed of ancient hedgerows and hedgerows about 100 years old (Baudry, 1985) show that some species, e.g. *Pyrus communis* and *Taxus baccata* are, at the least, slow colonizers as they are not found in the most recent hedgerows.

The 25 bird species found within the study area are distributed according to the complexity of the vegetation structure within landscape elements and not according to landscape structure. This is in agreement with other studies that show that hedgerow structure plays a major role for bird species distribution (Arnold, 1983; Osborne, 1984; Shalaway, 1985). They differ from the other groups in the size of their home range which is obviously small for forest carabids (Stork, 1990), but bigger for passerine birds (Blondel, 1979). The home range determines the scale at which species perceive their environment, for example, a landscape heterogeneous for beetles may be homogeneous for birds. In the study area all the hedgerows may easily be reached by the different bird species, which do not perceive any heterogeneity at the landscape scale.

Corridor efficiency and hedgerow structure

Hedgerows permit either just dispersion or also reproduction for some forest carabid beetles, which we called 'corridor' species (Burel, 1989). These beetles are not present in all the hedgerows of the network and their abundance depends on vegetation structure. The most favourable conditions are the presence of two parallel hedgerows bordering a lane, or a dense herbaceous layer and the presence of tree cover for a single hedgerow. This result is in agreement with other studies on corridors. For small forest mammals which use hedgerows as dispersal corridors (Fahrig and Merriam, 1985) the quality of the corridor is very important for the survival of the metapopulation (Henein and Merriam, 1990; Merriam and Lanoué, 1990). The presence of forest herbaceous species in shelterbelts or colonizing from woodlots, is also closely related to the width of the shelterbelts (Baudry and Forman, 1983; Baudry, 1988).

Lanes bordered by two hedgerows were once frequent landscape features, but, at least in western France, one hedgerow is often removed to enlarge the lane. These green lanes are little studied though Richards (1928) found that they harbour forest bryophytes. They can be considered as wide corridors and, when the canopy layer is continuous, a forest-like microclimate results, at least at ground level. Our carabid studies demonstrate that they are good habitats for forest species (Burel, 1989) and casual observations on plants yield similar results. The ground flora is different inside the lane compared to the field side, with many more forest interior species (Baudry, 1985). They certainly contain valuable reservoirs and deserve more study.

Populations in hedgerow networks: the case of carabid forest species

Forest 'corridor' carabid beetles, such as *Abax ater* are found in hedgerows, woodlots and forests. Following the metapopulation concept we hypothesized that small local populations, localized in network nodes such as lane intersections or small woodlots attached to hedgerows, are linked by hedgerows which the insects use as corridors for dispersal, and eventually for reproduction. Petit (Petit and Burel, 1994) showed by a capture-recapture experiment that *A. ater* is able to reproduce in the nodes of the network and that movements within these small landscape elements are similar to those found in the forest (Drach and Cancela da Fonseca, 1990). Preliminary investigations of individual movements in a wide corridor linking two small woodlots confirm the presence of local populations in nodes. Emigration from these populations is low. In the corridor itself the population density is lower, 0.3 individuals per m² compared to 5 in the wood (density computed according to the Jolly-Seberg method (Southwood, 1978)). Insect movements along the corridor are more important than in the nodes, the diffusion coefficient (Drach and Cancela da Fonseca, 1990) along the corridor being 143 m² per week, compared to 38 m² per week in the forest or in woodlots. Directional movements along hedgerows may be identified, while no distinct movement pattern exists for populations installed in woodlots (Petit, 1992). These preliminary results are consistent with a metapopulation model (Figure 6.1). Some reproduction does occurr within hedgerows, but these may be considered as suboptimal habitats.

Figure 6.1 Model of carabid forest 'corridor' population in a hedgerow network landscape

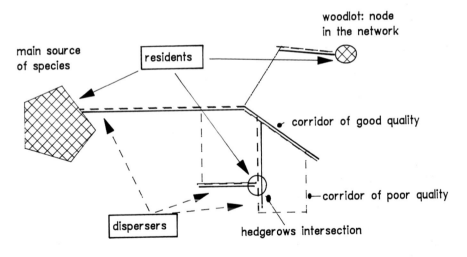

For *A. ater* at least, connectivity among small woody landscape elements is necessary to maintain its populations at the landscape level, as individuals disperse from local populations using only hedgerows. Otherwise, isolated local populations would become extinct, due to demographic or environmental stochastic events. This has been shown in rapidly changing landscapes, where progressive removal of hedgerows, associated with a sharp decrease in connectivity, leads to a decline in abundance of *Abax ater* in hedgerows (Burel, 1992).

Factors governing colonization and extinction processes

The relationship between landscape spatial patterns and species distribution patterns is not always at equilibrium and, when landscape changes, populations may not react synchronously. For example, there is a time-lag between hedgerow removal and extinction of isolated populations of *Abax ater* in other hedgerows (Burel, 1992). In a rapidly changing landscape there is evidence that changes vary in time and space (Burel and Baudry, 1990*a*) In a 2,000 hectare area where hedgerow removal had taken place between 1952 and 1985 'corridor' forest species remained abundant in hectare quadrats which were densely hedged in 1952, whatever the density in 1985. If only a few hedgerows are left, carabids find refuge in them, their abundance being abnormally high locally. This may be the result of a supersaturation effect, which appears when the availability of favourable habitats decreases rapidly (Clavreul, 1984; Blondel, 1986). Elsewhere, in quadrats where the network density was low in 1952 and remains low, the abundance of forest 'corridor' species is low in similar isolated woody elements (Burel, 1992). After a while, a relaxation phenomenon should occur and landscape and species spatial patterns will again be related to each other. For a while, carabid assemblages are a memory of the previous landscape. In this case, the time-lag is due to extinction taking time. However, on the other hand, when new habitats are created there can be a delay in their colonization by slow dispersers, even if the habitats are suitable (Burel, 1994). This may be illustrated by a decrease in abundance as one goes away from a source. In New Jersey, a 50-year-old hedgerow network was gradually colonized by herbaceous plant species such as *Impatiens capensis, Circaea quadrisulcata* and *Geum canadense*, which are mostly found in the remnant woodlots. The process is slow and there is still a strong effect of the distance from the relict woodlots that serve as sources of propagules (Baudry, 1985, 1988) (Figure 6.2).

Figure 6.2   Abundance of forest plant species in hedgerows with distance to the forest edge, in New Jersey

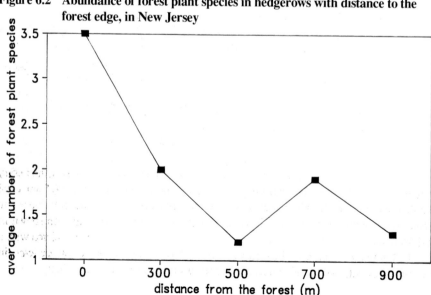

Hedgerows as sources of biotic effects on the surrounding elements

In the managed agricultural landscapes of western France hedgerows are often the only woody elements. When land abandonment occurs new habitats become available for forest species. A study in Pays d'Auge, Normandy, showed that carabid forest 'corridor' species colonize bramble patches from hedgerows (Burel and Baudry, 1994). As in hedgerow networks spatial connectivity among suitable elements is necessary for these species to disperse within a landscape. Here connection may be measured by the proximity between one patch and the nearest potential source element. Abundance of forest species in one patch depends on distance from hedgerows and size of patch which is, in turn, determined by its age. The role of hedgerows as sources of propagules has also been shown for plant species invading bramble patches in abandoned grassland (Burel and Baudry, 1990*b*). The presence of hedgerows accelerates sucessional processes and only species present in adjacent hedgerows are found in abandoned pastures. The successional patches are new nodes that participate in the maintenance of biodiversity in the landscape.

Conclusion

Agricultural landscapes are mainly driven by human activities and, contrary to natural ones (Shugart, 1984; Smith and Urban, 1988), their dynamics is not cyclic, and trajectories of change are complex and unpredictable at the landscape scale (Burel and Baudry, 1990*a*). Hedgerows are subjected to disturbance (e.g. pollarding), and local extinction of species may occur either because of stochastic environmental events or demographic events. Thus, species conservation in hedgerows must consider the possibilities of recolonization of existing hedgerows and/or colonization of new ones. Colonization is a function of the regional species pool and of species dispersal ability. The latter is often constrained by landscape structure.

Research carried out within the framework of landscape ecology emphasizes that hedgerows cannot be considered as isolated landscape elements. Hedgerow species diversity can only be understood as resulting from landscape scale processes. If management practices at hedgerow scale are important in order to maintain habitat quality, then landscape design and management to ensure a high connectedness within the network as well as connections with sources of forest species are major influences on species colonization of available habitats.

In countries where land reallotment programmes take place (e.g. France, Germany, The Netherlands, Belgium), removal and planting of hedgerows must be part of the new landscape design. To maintain or to restore network connectedness is the first goal. This does not necessarily conflict with efficient working conditions for farmers (Baudry and Burel, 1984). One might also expect that new hedgerows connected with older ones will have a faster increase of species diversity, though, to our knowledge, this has not been demonstrated.

The landscape perspective has implications for the monitoring biodiversity. First, it is preferable to monitor several connected hedgerows and to see how they change relative to each other. Second, time-lags between landscape changes (either plantation or hedgerow removal) and ecological pattern changes must be considered in data analysis. Third, spatial correlation in species distribution implies that classic parametric statistics cannot be used.

The practical goal of species conservation in hedgerows offers opportunities to test new ecological theories (metapopulations, connectivity in landscapes) and offers perspectives for research in conservation biology in agricultural landscapes. At the landscape level, processes of colonization, effects of cyclic disturbance and the role of field size merit more research to permit the formulation of specific guidelines for conservation.

Acknowledgement

We thank the Ministry of Environment (Comité EGPN) for supporting our research.

References

Agger P, Brandt J. (1988) Dynamics of small biotopes in Danish agricultural landscapes. *Landscape Ecology* **1**, 227-40.

Arnold GW. (1983) The influence of ditch and hedgerow structure, length of hedgerow, and area of woodland and garden on bird numbers on farmland. *Journal of Applied Ecology* **20**, 731-50.

Balent G, Courtiade B. (1992) Modelling landscape changes in a rural area of south-western France. *Landscape Ecology* **6**, 195-212.

Baudry J. (1985) Utilisation des concepts de landscape ecology pour l'analyse de l'espace rural: occupation du sol et bocage. Thèse de Doctorat d'Etat. Université de Rennes 1, 487pp.

Baudry J. (1988) Structure et fonctionnement écologique des paysages: cas des bocages. *Bulletin d'Ecologie* **19**, 523-30.

Baudry J, Burel F. (1984) Landscape project: Remembrement: Landscape consolidation in France. *Landscape Planning* **11**, 235-41.

Baudry J, Forman RTT. (1983) Hedgerows as corridors for forest plants in a New Jersey agricultural landscape. *Bulletin of the Ecological Society of America* **64**, 96.

Baudry J, Merriam HG. (1988) Connectivity and connectedness: functional versus structural patterns in landscapes. In: Shreiber KF, ed. *Connectivity in landscape ecology*. Proceedings of the 2nd international IALE seminar, Münstersche Geographische Arbeiten **29**, 23-8.

Bennett F. (1990) *Habitat corridors, their role in wildlife management and conservation*. Conservation and environment, Melbourne, Australia, 37pp.

Blondel J. (1979) *Biogéographie et écologie*. Masson. 221pp.

Blondel J. (1986) *Biogéographie évolutive*. Masson. 221pp.

Brandt J, Agger P. (eds) (1984) *Methodology in landscape ecological research and planning*, Vol. 5. Roskilde University Centre, pages 118, 150, 153, 171, 235.

Burel F. (1988) Biological patterns and structural patterns in agricultural landscapes. In: Schreiber KF, ed. *Connectivity in landscape ecology.* Proceedings of the 2nd international IALE seminar, Münstersche Geographische Arbeiten **29,** 107-10.

Burel F. (1989) Landscape structure effects on carabid beetles - spatial patterns in western France. *Landscape Ecology* **2,** 215-26.

Burel F. (1991) *Dynamique d'un paysage: réseaux et flux biologiques.* Editions du Muséum National d'Histoire Naturelle, Paris. 236pp.

Burel F. (1992) Effect of landscape structure and dynamics on carabids biodiversity, in Brittany France. *Landscape Ecology* **6,** 161-74.

Burel F. (1993) Time-lags between spatial pattern changes and distribution changes in dynamic landscapes. *Landscape and Urban Planning* **24,** 161-66.

Burel F, Baudry J. (1990*a*) Structural dynamics of a hedgerow network landscape in Brittany France. *Landscape Ecology* **4,** 197-210.

Burel F, Baudry J. (1990*b*) Hedgerow networks as habitats for colonization of abandoned agricultural land. In: Bunce RHG, Howard DC, eds. *Species dispersal in agricultural environments.* Belhaven Press, Lymington, 238-55.

Burel F, Baudry J. (1994) Reaction of ground beetles to vegetation changes following grassland dereliction. *Acta Oecologica* **15,** no.4.

Burgess RL, Sharpe DM. (eds) (1981) *Forest island dynamics in man dominated landscapes.* Ecological studies, **41.** Springer-Verlag, New York, Heidelberg, Berlin. 310pp.

Cameron RAD, Down K, Pannett DJ. (1980) Historical and environmental influences on hedgerow snail faunas. *Biological Journal of the Linnean Society* **13,** 75-87.

Clavreul D. (1984) Contribution à l'étude des interrelations paysage/peuplements faunistiques en region de grande culture. Thèse de doctorat de 3ième cycle, Université de Rennes 1. 317pp.

Den Boer PJ. (1981) On the survival of populations in a heterogeneous and variable environment. *Oecologia* **50,** 39-53.

Drach A, Cancela da Fonseca JP. (1990) Approche expérimentale des déplacements de Carabiques forestiers. *Revue d' Ecologie et Biologie du Sol* **27,** 61-71.

Fahrig L, Merriam HG. (1985) Habitat patch connectivity and population survival. *Ecology* **66,** 1762-8.

Forman RTT. (1983) Corridors in a landscape: their ecological structure and function. *Ekology CSSR* **2,** 375-87.

Forman RTT, Baudry J. (1984) Hedgerows and hedgerow networks in landscape ecology. *Environmental Management* **8,** 499-510.

Forman RTT, Godron M. (1986) *Landscape ecology.* John Wiley and Sons, 619pp.

Giot PR, L'Helgouach J, Monnier JL. (1979) *Préhistoire de la Bretagne.* Ouest France, 444pp.

Guyot G, Verbrugghe M. (1976) Etude de la variabilité spatiale du microclimat à l'échelle parcellaire en zone bocagère. In: *Les bocages: histoire, écologie, économie.* INRA, CNRS, ENSA et Université de Rennes, 131-6.

Hanski I. (1989) Metapopulation dynamics: does it help to have more of the same? *Trends in Ecology and Environment* **4,** 113-14.

Helliwell DR. (1975) The distribution of woodland plant species in some Shropshire hedgerows. *Biological Conservation* **7**, 61-72.

Henein K, Merriam G. (1990) The elements of connectivity where corridor quality is variable. *Landscape Ecology* **4**, 157-71.

INRA, ENSA et Université de Rennes (1976) *Les bocages: histoire, ecologie, economie.*

Lack PC. (1988) Hedge intersections and breeding bird distribution in farmland. *Bird Study* **35**, 133-6.

Lefeuvre JC, Missonnier J, Robert Y. (1976) Caractérisation zoologique. Ecologie animale (des bocages), Rapport de synthèse In: *Les bocages: histoire, ecologie, economie.* INRA, CNRS, ENSA et Université de Rennes, 315-26.

Leonard PL, Cobham RO. (1977) The farming landscape of England and Wales: a changing scene. *Landscape Planning* **4**, 205-36.

Levins R. (1970) Extinctions. In: *Some mathematical questions in biology, Lectures on mathematics in the life sciences,* Vol. **2**. American Mathematics Society, Providence, Rhode Island, 77-107.

Lewis T. (1969) The diversity of insect fauna in a hedgerow and neighbouring fields. *Journal of Applied Ecology* **6**, 453-8.

Mader HJ. (1988) The significance of paved agricultural roads as barriers to ground dwelling arthropds. In: Schreiber KF, ed. *Connectivity in landscape ecology.* Proceedings of the 2nd international IALE seminar, Münstersche Geographische Arbeiten **29**, 97-100.

Merriam HG. (1990) Ecological processes in the time and space of farmland mosaic. In: Zonneveld IS, Forman RTT, eds. *Changing landscapes: an ecological perspective.* Springer-Verlag, 121-33.

Merriam G, Lanoue A. (1990) Corridor use by small mammals: field measurement for 3 types of *Peromyscus leucopus. Landscape Ecology* **4**, 123-33.

Meyer J. (1972) L'évolution des idées sur le bocage en Bretagne. In: *La pensée géographique française contemporaine.* Presses universitaires de Bretagne, 453-67.

Milne BT, Forman RTT. (1986) Peninsulas in Maine: woody plant diversity, distance, and environmental patterns. *Ecology* **67**, 967-74.

Naiman RJ, Décamps H. (eds) (1990) *The ecology and management of aquatic-terrestrial ecotones.* Man and the biosphere, Series 4, UNESCO, Parthenon Publishing, 316pp.

Opdam P. (1991) Metapopulation theory and habitat fragmentation: a review of holarctic breeding bird studies. *Landscape Ecology* **5**, 93-106.

Osborne P. (1984) Bird numbers and habitat characteristics in farmland hedgerows. *Journal of Applied Ecology* **21**, 63-82.

Peterken GF, Game M. (1981) Historical factors affecting the distribution of *Mercurialis perennis* in Central Lincolnshire. *Journal of Ecology* **69**, 781-96.

Petit S. (1992) Diffusion of forest carabid beetles in hedgerow network landscapes. *8th European carabidologists' meeting. 1-4 Sept.,* Louvain la Neuve, Belgique.

Petit S, Burel F. (1993) Movement of *Abax ater* (Col. Carabidae): do forest species survive in hedgerow networks. *Vie et Milieu* **42**, 119-24.

Pollard E. (1968) Hedges. III. The effect of removal of the bottom flora of a hawthorn hedgerow on the Carabidae of the hedge bottom. *Journal of Applied Ecology* **5,** 125-39.

Pollard E, Hooper MD, Moore NW. (1974) *Hedges.* London: Collins. 256pp.

Richards PWN. (1928) Ecological notes on the bryophytes of Middlesex. *Journal of Ecology* **16,** 267-300.

Saunders DA, Hobbs RJ. (1991) *The role of corridors.* Surry Beaty & Sons, 442pp.

Shalaway SD. (1985) Fencerow management for nesting birds in Michigan. *Wildlife Society Bulletin* **13,** 302-6.

Schreiber KF. (ed.) (1988) *Connectivity in landscape ecology.* Proceedings of the 2nd international IALE seminar, Münstersche Geographische Arbeiten **29.** 255pp.

Shugart H. (1984) *A theory of forest dynamics.* Springer-Verlag. 278pp.

Smith TM, Urban DL. (1988) Scale and resolution of forest structural pattern. *Vegetatio* **74,** 143-50.

Soltner D. (1978) *L'arbre et la haie.* Collection sciences et techniques agricoles, 104pp.

Southwood TRE. (1962) Migration of terrestrial arthropods in relation to habitat. *Biological Reviews* **37,** 171-214.

Southwood TRE. (1978) *Ecological methods with particular reference to the study of insect populations.* Chapman and Hall. 524pp.

Stork NE. (ed.) (1990) *Ground beetles: their role in ecological and environmental studies.* Intercept, Andover.

Thiele HU. (1977) *Carabid beetles in their environments.* Springer-Verlag, Berlin, Heidelberg, New York. 39pp.

Tjallingii SP, de Veer AA. (eds) (1981) *Perspectives in landscape ecology.* Proceedings of the International Congress Veldhoven, The Netherlands, Centre for Agricultural Publishing and Doc Wageningen. 344pp.

Wegner F, Merriam G. (1979) Movements by birds and small mammals between a wood and adjoining farmland habitats. *Journal of Applied Ecology* **16,** 349-58.

Yahner RH. (1983) Seasonal dynamics, habitat relationships, and management of avifauna in farmstead shelterbelts. *Journal of Wildlife Management* **47,** 85-104.

CHAPTER 7

MODELS RELATING BIRD SPECIES DIVERSITY AND ABUNDANCE TO FIELD BOUNDARY CHARACTERISTICS

T. Parish, T. H. Sparks and K. H. Lakhani
NERC Institute of Terrestrial Ecology, Monks Wood, Abbots Ripton, Huntingdon, Cambridgeshire PE17 2LS, UK

Introduction

In the agricultural lowlands of Great Britain, the hedgerow is frequently the most important landscape feature. It is also an important element of the countryside as a habitat for wildlife.

The field boundary may be the only remnant of semi-natural habitat within a locality and its structure and management will therefore have a big effect on its conservation value. The value of a particular field boundary to birds depends entirely upon the management of the boundary and adjacent crops and its position relative to other habitats in the area. Thus the value of the 'best' parts of the agricultural landscape for birds is in the hands of the farmer. Birds are relatively mobile animals and are generally near the top of the food chain. They are influenced by the relative abundance of organisms lower in the chain. Their ability to select home ranges which are optimal in terms of food, shelter, nest sites etc. is reflected in their observed distributions.

Review of birds in hedgerows

Bird surveys during the late 1920s and 1930s on the former Oxford University Farm (Alexander, 1932; Chapman, 1939) suggested that bird densities in farmland depended much more on the mileage and thickness of hedgerows than on the acreage of land between them. Alexander (1932) reported that the most consistently observed farmland birds were blackbird, chaffinch, robin, song thrush, partridge, wren, starling, and blue tit.

Chapman (1939) reported the results of extended surveys to include arable fields as well as pasture. Chaffinch, house sparrow, linnet, skylark, lapwing, and partridge appeared more numerous in the arable part of the survey. Conversely, starling, yellow hammer, blue tit, fieldfare, redwing, blackbird, song thrush, and robin appeared more numerous in the pasture part of the survey. Chapman (1939) also reported a greater density of birds on arable land, mainly due to flocks of finches and skylarks. He concluded that, despite numbers being higher in pasture, non-flocking bird species were equally numerous on pasture and arable land when considered per mile of hedgerow -

the arable portion having 1.2 miles of hedge per 100 acres compared to 3.1 miles per 100 acres in the pasture.

More recent work has emphasized that the majority of birds in farmland rely on trees and hedgerows (Wyllie, 1976) and that fewer birds are observed in more intensively managed hedges (Williamson, 1971). The shrub species composition of the hedgerow appears to be important in determining bird abundance, with diverse hedgerows and hawthorn being more valuable (Arnold, 1983; Osborne, 1984).

There is a large amount of literature on other aspects of hedgerow birds and these include the works of Hooper (1970), Pollard *et al.* (1974), Murton and Westwood (1974), Lack (1990), and Peay (1990). O'Connor and Shrubb (1986) have reviewed the relationship between farming and birds and the reader is recommended to refer to this text for further information.

In order to establish the relative merits of field boundaries and the influence of the adjoining crop the Institute of Terrestrial Ecology (ITE) has undertaken two projects to study bird populations. The studied field boundaries included those with no hedges and those with hedges of varying sizes. A range of size and management of other linear components such as verges and ditches were also present. The first study, at a number of sites in the Huntingdon area, assessed the effects on birds of hedgerow and other field boundary attributes as well as the influence of the adjacent crop (arable or pasture) and the effect of field size.

The second study, from which material for this paper is extracted, extended the work by looking at bird distribution across a catchment of a river tributary at Swavesey, Cambridgeshire. In addition to ongoing typical farming systems, this included an area undergoing rapid agricultural change following the installation of improved drainage. There was an opportunity to confirm the earlier findings about the influence of hedge size and adjacent crop on bird distribution. This was also a unique opportunity to study the effects upon bird populations of a rapidly changing drainage regime, changing farming practices and crop husbandry, and consequent changes in field boundary management.

Materials and methods

The Swavesey fens lie on the flood plain of the Great Ouse in Cambridgeshire. At this point the river is some 3 metres above sea level and the study area is part of the catchment of one tributary and lies entirely below the 20 metre contour. The area contains no woods. From 1985 almost half of the flood meadows forming the Swavesey fens were included in a new pump drainage and flood protection scheme. A major study of the hydrological and ecological consequences of this change was instigated by ITE and the Soil and Water Research Centre (SWRC, formerly the Field Drainage Experimental Unit) of the Agricultural Development Advisory Service (ADAS). Ecological monitoring involved surveys of ditch, bank and field flora, invertebrates, aquatic fauna, and birds. As well as the area affected by improved drainage the study incorporated, as controls, areas of traditional flood meadow, a conservation area and an area of intensive agriculture.

Figure 7.1   Map of the Swavesey fens

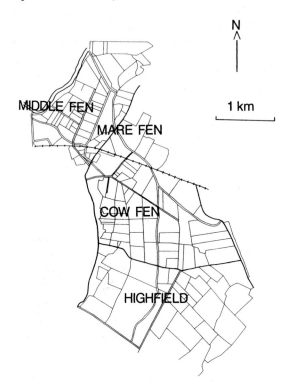

Between 1985 and 1991 bird surveys were carried out in three winter and three summer seasons. The surveys concentrated on the bird populations of field boundaries, whether these included a hedge or otherwise. This paper reports on the results of these six surveys, the differences between boundary types, the differences between adjacent land uses and the changes taking place over time. The study described here covered four distinct areas (Figure 7.1). *Middle Fen* is an area of predominantly flood meadow pasture, close to the river and subjected to annual flooding, which witnessed some unsuccessful arable conversion during the second half of the nineteenth century and during the Second World War. Middle Fen contained a large proportion of tall hedges (Figure 7.2a), although some of these were no longer continuous, with many of the gaps resulting from the death of hawthorn following waterlogging. *Mare Fen* is an area managed by the local naturalist's trust as a flood meadow, with an extended period of controlled winter flooding and water levels kept artificially high in the ditches during the summer. As it consists of a single field of 17 hectares, it contained fewer transects than the other areas (Figure 7.2b). *Cow Fen* is similar historically to Middle Fen in terms of land use, landscape and flooding, but has been protected by the new pump drainage scheme since 1985. Traditionally, land use had been focused on grazing and hay-making, but following the provision of improved drainage and flood control there has been a rapid conversion to arable, together with changes in field boundary management.

In 1985 a range of hedgerow heights (Figure 7.2c) were present, although many lengths of hedgerow had already been removed to enlarge fields for conversion to arable or in the creation of the new network of drainage channels. By 1991 (Figure 7.2d) there had been a further movement towards shorter, more frequently managed hedgerows. Finally, at *Highfield*, an area to the south of the study area, the fields are gravity drained and traditionally intensive arable and mixed farming. Short hedgerows were the only ones remaining over much of the Highfield landscape (Figure 7.2e).

Figure 7.2   The distribution of hedgerow heights in the study area
(a) Middle Fen 1985

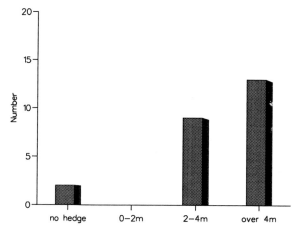

(b) Mare Fen 1985

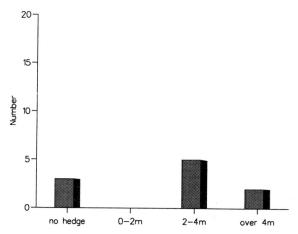

(c) Cow Fen 1985

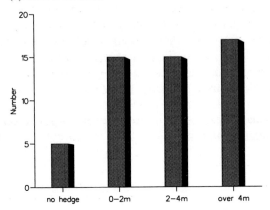

(d) Cow Fen 1991

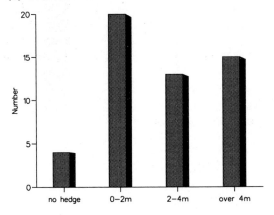

(e) Highfield 1985

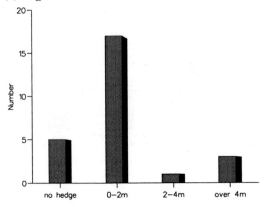

Arable farming in the Swavesey area consists predominantly of cereal crops on the heavy clay soils, although some beans and oilseed rape are also grown. Dairy cattle were the most common domestic grazing animals, with some beef cattle, a flock of sheep and horses in the smaller fields close to the village. Hedges, other than parish boundaries and those enclosing small fields near to the village ('closes'), were established at the time of Enclosure (around 1834) and were predominantly hawthorn. Because of the nature of farming in the Swavesey fens they played an important role as stockproof boundaries, a role no longer essential in continuous arable regimes. The major ditches in the study area have an important drainage function and were wide and deep as a consequence of regular dredging and vegetation clearance. The main drainage artery (Uttons Drain and Swavesey Drain) also received a major input from a sewage treatment works as well as a large agricultural catchment.

The bird surveys consisted of 131 field boundary transects spread across the study area (Figure 7.3). These transects were each 200 metres long and extended 10 metres into the crop on either side. The majority of these field boundaries contained hedges ranging from intensively managed to unmanaged and some incorporated major arterial drains. Surveys were undertaken in the winters of 1985/86, 1986/87 and 1991/92 and the summers of 1986, 1987 and 1991. In each season each transect was recorded at approximately monthly intervals. At the beginning of each transect the boundary was inspected with binoculars. The transect was then slowly walked and the numbers of birds seen or heard within the transect area and the presence of occupied nests were recorded. Birds were identified to individual species.

Figure 7.3   The location of transects in the study area

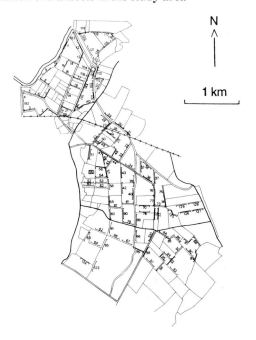

In addition to the records of bird populations the flora and structure of the field boundary was recorded. Many of the hedges had lost their stock-proof function and were gappy or survived only as a line of scattered shrubs. Trees (generally ash or willow, with oak and elm less common) were present in some transects. Small hedges tended to be trimmed annually, often severely. The large unmanaged hedges had been left uncut for many years, some for more than 70 years. Hedges were dominated by hawthorn, with bramble and dog rose forming an understorey in larger hedges. Blackthorn, elm, buckthorn, and field maple were also present in some field boundaries. The physical characteristics of the transect were recorded, including hedge, verge, tree, and ditch dimensions together with a botanical survey of the transect flora and a record of the adjacent crops. For land use purposes, boundaries were classified as three types:

- grass, permanent or rotational grassland on both sides;
- arable, arable crops on both sides;
- intermediate, grass on one side and arable on the other.

Statistical modelling

The major variables of interest in this study were the numbers of species. This has been expressed in three ways as total number of species per transect, mean number of species per transect per visit and number of nesting bird species per transect. Simpson's index of diversity (Simpson, 1949), incorporating both number of species and species abundance in each transect, was calculated for each of the surveys. There were 35 species for which there was sufficient data (usually indicated by presence in at least 10 transects) to justify modelling the abundance of individual bird species. Seasonal migrants were modelled either for the summer seasons (e.g. willow warbler) or for the winter (e.g. fieldfare).

Species numbers and abundance variables and Simpson's index were examined in relation to the physical characteristics of the field boundary and the land use of the adjacent fields. Potential explanatory variables and factors are listed in Table 7.1.

All modelling of Swavesey data was carried out using the GENSTAT statistical package. The approach used was essentially standard multiple regression. Crop (adjacent land use) was included in all models as a qualitative factor and separate slopes were fitted for each crop (i.e. interactions with crop were included). The importance of an explanatory variable was assessed in each survey, but its consistency over surveys was critical in whether it was selected for the final model; for each dependent variable, models for each survey used the same set of explanatory variables. The final models thus obtained necessarily had slightly lower R^2 values than models selected purely on statistical arguments. Models selected under this approach were statistically significant and biologically robust.

Table 7.1 Potential explanatory variables in modelling bird species numbers and abundance

Tree height	Verge width
Tree number	Ditch depth
Product of tree height and number	Ditch width
Hedge length	Product of ditch width and depth
Hedge height	Combined widths of ditch, verge and hedge base
Hedge crown width	Hedge height to crown width ratio
Hedge base width	Number of aquatic plant species
Product of hedge length and width	Number of herbaceous plant species
Triple product of hedge length, width and	Number of woody plant species
Base width	Total number of plant species
Triple product of hedge length, width and	Percentage grass cover
Crown width	Percentage bare ground
Crop 1=pasture, 2=intermediate, 3=arable	

Results

A total of 77 bird species were observed in the surveys. Some of these species were only observed rarely. Appendix 7.1 lists the recorded bird species and gives an indication of their relative abundance: only selected results are presented in this paper.

Table 7.2 summarizes the results from these regression models. For each species or species numbers category the range of R^2 values is given together with an indication of the levels of statistical significance achieved.

A total of 206 models were investigated. Statistical significance was achieved in all but 44 of these and the majority were significant at the 0.1% level of significance. Model significance varied from species to species; for example, blackbird and robin data produced consistently good models whereas pheasant and particularly yellow wagtail data produced poor models. Non-significant models were sometimes, but not always, associated with a scarcity of data.

Table 7.2   Summary of the range of R^2 values from the six bird surveys, and an indication of the significance levels achieved

	Range of R^2	Number of models at given significance level			
		ns	5%	1%	0.1%
Species numbers					
Total/transect	56-66%	0	0	0	6
Mean/transect/visit	60-72%	0	0	0	6
Nesting/transect	68-70%	0	0	0	3 (summer only)
Simpson's index	27-38%	0	0	0	6
Abundance					
Mallard	20-27%	0	0	3	3
French partridge	3-19%	2	2	1	1
Pheasant	2-17%	4	0	1	1
Moorhen	14-42%	2	0	0	4
Snipe	2-23%	4	0	1	1
Wood pigeon	10-56%	1	0	1	4
Skylark	10-34%	1	0	0	5
Carrion crow	12-46%	1	0	0	5
Magpie	12-51%	1	1	0	4
Great tit	13-42%	1	0	0	5
Blue tit	25-49%	0	0	0	6
Wren	18-42%	0	0	1	5
Song thrush	8-33%	1	1	0	4
Blackbird	48-60%	0	0	0	6
Robin	37-56%	0	0	0	6
Dunnock	14-47%	0	1	1	4
Meadow pipit	17-26%	0	0	0	6
Starling	2-23%	4	0	1	1
Greenfinch	10-31%	2	1	2	1
Goldfinch	5-53%	1	0	1	4
Linnet	2-31%	2	1	0	3
Bullfinch	23-38%	0	0	1	5
Chaffinch	21-46%	0	0	0	6
Yellowhammer	2-55%	3	1	1	1
Reed bunting	8-39%	2	0	0	4
House sparrow	9-27%	1	2	0	3
Heron	4-28%	1	1	0	1 (3 only)
Turtle dove	13-34%	0	0	2	1 (3 only)
Sedge warbler	13-26%	0	1	1	1 (3 only)
Corn bunting	5-21%	0	1	0	2 (3 only)
Willow warbler	23-38%	0	0	0	3 (3 only)
Whitethroat	11-24%	1	0	1	1 (3 only)
Yellow wagtail	-	6	0	0	0
Fieldfare	8-26%	1	1	0	1 (3 only)
Redwing	9-11%	2	0	0	0 (2 only)
		44	**14**	**19**	**129**
				total (206)	

Hedge height

In the following comments short hedges are taken to be those less than 2 metres in height, medium hedges those between 2 and 4 metres, and tall hedges those over 4 metres. Bird species numbers depended heavily on hedgerow size (Figure 7.4) with approximately twice the number of species occurring in tall hedges compared with short hedges. Indeed species numbers in short hedges were slightly lower than in those field boundaries which contained no hedges. These hedgeless transects often contained a wide grass verge or a large drain providing more habitat opportunities. This pattern was repeated for mean species numbers and for nesting species numbers, but the discrepancy between short and tall hedges was even greater for nesting species (Figure 7.4). Combining species numbers and abundance into Simpson's index of diversity (Figure 7.5) again emphasized the importance of hedgerow size on bird populations.

Figure 7.4   Bird species numbers by hedgerow height category (total numbers averaged over six seasons, nesting numbers averaged over three summer surveys)

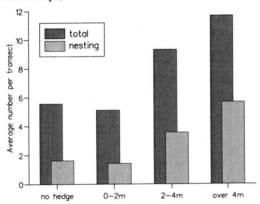

Figure 7.5   Simpson's index of diversity by hedgerow height category (averaged over six seasons)

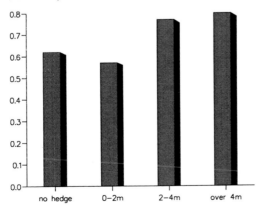

An examination of individual species revealed information on hedgerow preferences. Some, such as skylark and the game birds, appeared to prefer the bare (i.e. predominantly arable) field boundaries. Others, such as meadow pipit, seemed unaffected by hedgerow size. Conversely, a large number of species, for example, robin (Figure 7.6), blackbird (Figure 7.7) and wren (Figure 7.8) were much more abundant in larger hedges, the short hedges failing to provide the required type of habitat.

Figure 7.6   Robin abundance by hedgerow height category (averaged over six seasons)

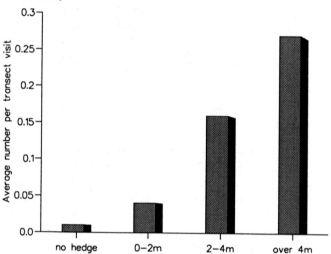

Figure 7.7   Blackbird abundance by hedgerow height category (averaged over six seasons)

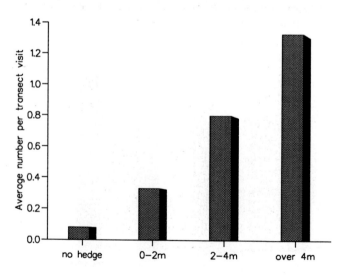

Figure 7.8   Wren abundance by hedgerow height category (averaged over six seasons)

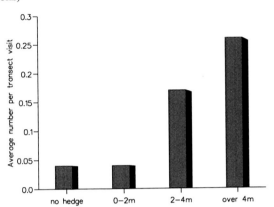

Land use

Species numbers were greater in grass boundaries than in arable boundaries and, as might be expected from the above finding, intermediate in the mixed boundaries (Figure 7.9). This was apparent in both winter and summer and for nesting birds and Simpson's index (Figure 7.10). Predictably, the intermediate category fell between grass and arable in terms of species numbers. Such differences were apparent above and beyond the differences in hedgerow dimensions that existed between the land use categories. Again a number of birds showed a predilection for one or other land use. Robins and a number of other 'woodland' birds showed a clear preference for grass transects whilst skylark and corn bunting showed a reverse pattern.

Figure 7.9   Bird species numbers by land use category (each of summer, winter and nesting are averaged over three surveys)

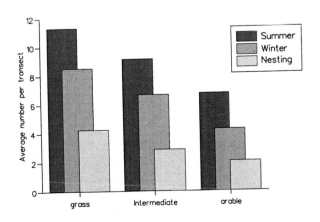

Figure 7.10   Simpson's index of diversity by land use category (av. of six surveys)

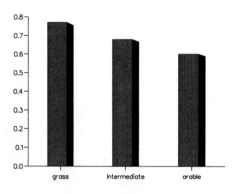

Fens

The differences between species numbers and abundance in the separate fens was partly, but not wholly, influenced by the hedgerow dimensions and land use. In Middle Fen large hedges dominated whilst in Highfield small hedges were most numerous. Cow Fen and Mare Fen were intermediate between the two. As might be expected species numbers were greatest in the flood meadow environment of Middle Fen and lowest in the intensive farming area of Highfield (Figure 7.16).

Changes over time

During the period 1985 to 1991 there was a dramatic decline in the number of birds in the Swavesey fens. This decline coincided with a period of prolonged drought (Figure 7.11). In spite of this, some species have reversed the decline and pheasant, wood pigeon and magpie seemed to be on the increase (Figure 7.12). More typically, however, there was a decline in numbers, this appearing most serious for snipe, song thrush, blackbird, greenfinch, starling, and linnet (Figure 7.13).

Figure 7.11   Rainfall at ADAS Boxworth 1985-1991

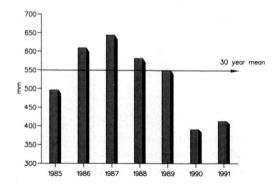

Figure 7.12   Bird species showing an increase over the period 1986-1991

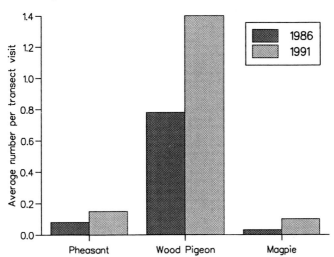

Figure 7.13   Bird species showing a decline over the period 1986-1991

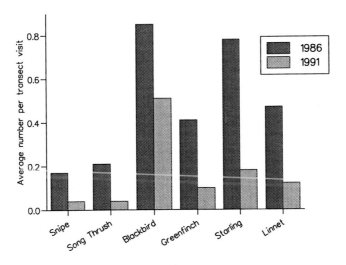

Figure 7.14 shows a decline in species numbers irrespective of hedgerow height, although this appeared most marked where no hedgerow existed. The decline in species numbers was similar in all hedgerow categories, but the *proportional* loss was obviously much greater in transects with a short hedge or no hedge.

Figure 7.14   Bird species numbers decline 1986-1991 by hedgerow height category

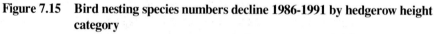

Figure 7.15 shows the decline in nesting species numbers. Again the decline appeared proportionally more serious for the 'short' and 'no hedge' transects. An examination of the changes in the fen categories is quite informative and was similar for both species numbers (Figure 7.16) and Simpson's index. Prior to the start of the study we would have expected Cow Fen to be similar in most respects to Middle Fen. Since drainage and flood control have been introduced there has been a conversion in Cow Fen to more intensive agriculture and a reduction in some hedgerows. Already we are witnessing the landscape and land use of Cow Fen changing towards that of Highfield. Decline in species numbers was most marked in Cow Fen, appearing to diverge away from Middle Fen and tend towards Highfield (Figure 7.16). A similar pattern was observed for Simpson's index except that the index for Cow Fen in 1991 dropped to that of Highfield. Mare Fen is a managed conservation area and one might hope that its species complement would approach that of Middle Fen, but besides being a single field of 17 hectares and rather devoid of hedgerows and trees, there have also been problems in achieving and maintaining the desired water levels.

Figure 7.15   Bird nesting species numbers decline 1986-1991 by hedgerow height category

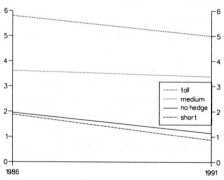

Figure 7.16   Bird species numbers decline 1986-1991 by fen

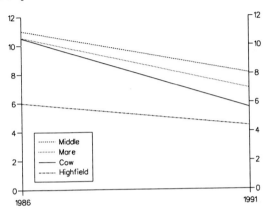

The effect of hedgerow trimming

In Cow Fen, following the drainage and flood control referred to earlier, there has been a rapid conversion to more intensive forms of agriculture. This has often been linked to a major reduction in the size of the field boundary hedgerows. Results from three boundaries, transects 35, 38 and 49, illustrate the effects of extensive pruning.

Transect 35: At the start of the study this treeless hedgerow was 4 metres tall and 3 metres wide at the crown. During 1990/91 its height and width were reduced to 2 metres and 1 metre, respectively. Originally a grass/arable boundary, cropping on both sides is now arable.

Transect 38: This hedgerow originally measured 5 metres tall and 3 metres wide at the crown. It contained, and still does contain, trees. During 1990/91 the hedge was reduced to 1.3 metres tall and 1 metre wide at the crown. The grass/arable land use at the start of the study period has been changed to arable.

Transect 49: Originally 4.5 metres tall with a 3-metre-wide crown, this hedgerow was reduced in autumn 1986 to 1.5 metres tall and 1 metre wide. Its height has since been further reduced to 1 metre. It retains its single tree. Cropping in adjacent fields was originally mixed and remained so in 1991.

Total species numbers and nesting species numbers are documented in Table 7.3. Before trimming, these hedges contained an average, or larger, number of bird species compared to Cow Fen as a whole. Transect 38 had a particularly large complement, possibly a result of its slightly greater height and the additional presence of trees. Whilst species numbers declined throughout Cow Fen, and indeed throughout the study area, the decline appeared to be particularly severe in these transects. After size reduction there were no breeding birds recorded in these three transects and total species numbers had declined to less than half of the Cow Fen average. In times of low bird numbers it would appear that these heavily trimmed hedgerows are among the first to be abandoned as nest sites.

Table 7.3 Total and nesting species numbers for the three illustrative transects which experienced severe trimming together with the mean for Cow Fen as a whole; pruning occurred in the intervals indicated by a vertical line

	Species numbers category	Mean W85/S86	Mean W86/S87	Mean S91/W91
Transect 35	Total	9.0	7.5	2.5
	Nesting	3.0	3.0	0.0
Transect 38	Total	12.5	18.5	2.5
	Nesting	7.0	7.0	0.0
Transect 49	Total	9.5	5.5	2.0
	Nesting	3.0	0.0	0.0
All Cow Fen	Total	9.9	10.2	5.7
	Nesting	3.8	3.7	3.0

Discussion

Data from surveys, rather than designed experiments, often suffer from a lack of independence in the explanatory variables. Our surveys were not exceptions, for instance, arable fields tended to be associated with short, narrow hedges. In such situations it may not be possible definitively to ascribe cause and effect and in reaching conclusions from such data it is necessary to employ both scientific and subjective judgement.

The success of modelling individual bird species was variable. In some cases, such as blackbird, consistently good models were achieved. Others, such as French partridge, probably depended less on the attributes being modelled and the opposite was the case. In general, the models constructed from these surveys achieved high levels of statistical significance. Many of the models reinforced the known ecology of the species and helped quantify the effects of various landscape features.

The influence of hedge and tree dimensions on bird species numbers and the abundance of many individual bird species was very marked. A heavily trimmed hedge devoid of berries frequently supported fewer birds than a bare grassy verge or ditch with no hedge.

A surprisingly large number of species showed a clear decline in numbers from pasture through intermediate to arable transects. The reasons for this may include reduced food supply from insects, berries and other seeds and more human disturbance.

Conversely, there were a number of species, such as the skylark and corn bunting, adapted to an arable environment.

Hedgerows play a major part in the size and composition of bird populations. There were no woods in the Swavesey area and hedgerows provided the only habitat for the so-called tree or woodland species. The size and structure of the hedgerow were critical in determining the abundance of individual species and species numbers in general.

Management of field boundaries is expensive and generally the greater the inputs, the lower the value to wildlife. Typical inputs are:

- annual hedge trimming;
- flailing of verges, ditches and banks far beyond the level to ensure the drainage function of the ditch or to prevent scrub encroachment;
- the use of total herbicides as sterile strips round the crop, frequently sprayed on to verges or the hedge bottom in an attempt to prevent weeds encroaching into the crop, but in fact favouring annual weeds such as cleavers and bromes and thus reinforcing the need for further and more drastic herbicide treatments.

It has been argued by the Farming and Wildlife Advisory Group (FWAG) and others that attempting to grow a crop on the 1 or 2 metres adjacent to the boundary is not profitable considering the routine and additional special inputs required and the lower yield produced. There is, therefore, an option to create a perennial grass strip adjacent to the boundary and concurrently to allow the hedge to grow larger.

Under the recent EU Common Agricultural Policy for Set-aside it appears that area payments for arable crops and for Set-aside will use the Ordnance Survey area value for the whole field rather than the cultivated area. Consequently, there is less pressure on the farmer to maximize his cropped area by cultivating very close to the boundary. Grass verges and larger hedges could be established in the Set-aside year(s) and retained permanently. Grass verges replace the need for herbicide on the boundary and allow a clearly defined permanent division between the verge and the cultivated and cropped area. Set-aside not only offers opportunities to establish grass verges and allow hedge size to increase, but may, if maximum environmental benefits are eventually included in any long-term options, allow for the development of permanent pasture. This could be achieved by sowing an appropriate grass mix, or by natural regeneration and could involve whole fields. A greater benefit would be achieved by adopting marginal strips, say 15 metres wide, around many fields thus spreading the benefits more widely, while concentrating on the area adjacent to the field boundary. Together with the reduced need for hedgerow management, these measures would be expected to enhance wildlife generally, and bird populations in particular.

Conversion of short hedges in intensive arable to tall hedges with permanent pasture alongside might be expected to almost double species numbers. Planted trees could eventually further enhance this figure. Cutting regimes are an important management tool. Newly planted hedges should be trimmed annually such that annual increases in height or width are controlled. This will establish a dense growth form. Similar management should be applied to new growth after coppicing a poor hedge. Coppicing is usually more effective than simply allowing a badly managed short hedge to grow bigger.

Once boundary features are established, reduced frequency of cutting is not only beneficial to wildlife, but a significant cost saving. A two year rotation, e.g. alternate sides or lengths each year, is easy to implement for hedges, verges and ditch banks but longer rotations may have additional benefits in allowing bramble etc. to establish. Cutting can also be selective, such as leaving stems of woody species to grow to hedgerow trees or leaving bramble patches and species such as teasel and knapweed as sources of nectar or seeds for birds in winter.

Summary

The majority of birds in farmland depend on the field boundary and, more importantly, on the hedgerow to provide a viable habitat. This habitat is easily disturbed by man intent on increased productivity or neatness. Conversely, minimum or non-intervention treatments should allow even heavily trimmed hedgerows to provide refuges for increased bird numbers.

Different bird species have different habitat requirements and the creation of a variety of field boundaries in terms of their ditch, verge, tree and hedgerow structure are most likely to increase the diversity and abundance of the bird population.

Acknowledgements

The authors would like to thank the many farmers who allowed us regular access to their land and provided background information. We would also like to thank ADAS SWRC for their collaboration in the Swavesey fens project and ADAS Boxworth for the provision of rainfall data. This work was partly funded by The Ministry of Agriculture, Fisheries and Food.

References

Alexander WB. (1932) The bird population on an Oxfordshire farm. *Journal of Animal Ecology* **1**, 58-64.

Arnold GW. (1983) The influence of ditch and hedgerow structure, length of hedgerows, and area of woodland and garden on bird numbers in farmland. *Journal of Applied Ecology* **20**, 731-50.

Chapman WMM. (1939) The bird population on an Oxfordshire farm. *Journal of Animal Ecology* **8**, 286-99.

Hooper MD. (1970) Hedges and birds. *Birds* **3**, 114-17.

Lack PC. (1990) Farming and birds. In: *Fourth report of TERF, The Environment Research Fund: Farming and conservation management - putting new ideas into practice.* Inter-Regional Group, Norwich, 7-8.

Murton RK, Westwood NJ. (1974) Some effects of agricultural change on the English avifauna. *British Birds* **67**, 41-69.

O'Connor RJ, Shrubb M. (1986) *Farming and birds.* Cambridge University Press.

Osborne PJ. (1984) Bird numbers and habitat characteristics in farmland hedgerows. *Journal of Applied Ecology* **21**, 63-82.

Peay S. (1990) Managing field boundaries. In: *Fourth report of TERF, The Environment Research Fund: Farming and conservation management - putting new ideas into practice*. Inter-Regional Group, Norwich, 12-14.

Pollard E, Hooper MD, Moore NW. (1974) *Hedges*. London: Collins.

Simpson EH. (1949) Measurement of diversity. *Nature* **163**, 688.

Williamson K. (1971) A bird census study of a Dorset dairy farm. *Bird Study* **18**, 80-96.

Wyllie I. (1976) The bird community of an English parish. *Bird Study* **23**, 39-50.

Appendix 7.I   Mean number of birds recorded per transect per visit

	Winter			Summer		
	1985	1986	1991	1986	1987	1991
Little grebe	0.00	0.00	0.00	0.00	0.00	0.00
Heron	0.01	0.01	0.01	0.02	0.01	0.03
Mallard	0.20	0.08	0.14	0.20	0.22	0.19
Tufted duck	0.00	0.00	0.00	0.00	0.02	0.00
Mute swan	0.01	0.05	0.03	0.06	0.06	0.06
Kestrel	0.01	0.01	0.01	0.01	0.02	0.02
French partridge	0.07	0.20	0.10	0.11	0.13	0.08
Grey partridge	0.06	0.01	0.02	0.02	0.00	0.01
Pheasant	0.07	0.05	0.13	0.11	0.07	0.16
Moorhen	0.10	0.09	0.09	0.13	0.29	0.16
Coot	0.00	0.00	0.00	0.00	0.02	0.00
Lapwing	0.03	0.01	0.00	0.02	0.01	0.00
Golden plover	0.01	0.00	0.00	0.00	0.00	0.00
Snipe	0.17	0.31	0.04	0.03	0.03	0.05
Black-headed gull	0.00	0.00	0.00	0.00	0.00	0.00
Wood pigeon	0.91	0.46	1.33	1.11	1.24	1.46
Turtle dove	0.00	0.00	0.00	0.13	0.14	0.06
Collared dove	0.07	0.16	0.00	0.13	0.14	0.04
Little owl	0.00	0.00	0.00	0.02	0.02	0.02
Swift	0.00	0.00	0.00	0.00	0.01	0.00
Skylark	0.15	0.05	0.03	0.11	0.16	0.03
Carrion crow	0.10	0.13	0.08	0.14	0.13	0.16
Rook	0.07	0.11	0.02	0.13	0.03	0.05
Jackdaw	0.00	0.00	0.00	0.04	0.00	0.01
Magpie	0.04	0.05	0.12	0.01	0.03	0.09
Great tit	0.21	0.14	0.14	0.25	0.24	0.28
Blue tit	0.30	0.19	0.22	0.27	0.45	0.34
Long-tailed tit	0.06	0.03	0.06	0.07	0.09	0.07
Tree creeper	0.01	0.01	0.00	0.00	0.02	0.00
Wren	0.20	0.13	0.07	0.12	0.18	0.12
Fieldfare	0.10	0.82	0.23	0.00	0.00	0.00
Song thrush	0.23	0.22	0.03	0.21	0.12	0.05
Blackbird	0.72	1.05	0.64	0.66	0.74	0.40
Robin	0.13	0.13	0.16	0.13	0.11	0.12
Reed warbler	0.00	0.00	0.00	0.06	0.10	0.05
Sedge warbler	0.00	0.00	0.00	0.12	0.13	0.04
Dunnock	0.34	0.28	0.15	0.23	0.29	0.21
Meadow pipit	0.07	0.08	0.02	0.10	0.06	0.04
Pied wagtail	0.02	0.07	0.00	0.00	0.00	0.00

	Winter			Summer		
	1985	1986	1991	1986	1987	1991
Starling	0.66	0.69	0.20	0.87	0.67	0.17
Greenfinch	0.12	0.57	0.08	0.25	0.26	0.13
Goldfinch	0.22	0.23	0.07	1.05	0.74	0.38
Linnet	0.23	0.21	0.02	0.73	0.48	0.21
Bullfinch	0.20	0.15	0.08	0.09	0.11	0.06
Chaffinch	0.13	0.10	0.08	0.17	0.17	0.13
Yellowhammer	0.25	0.31	0.13	0.50	0.47	0.35
Corn bunting	0.01	0.00	0.00	0.07	0.07	0.03
Reed bunting	0.14	0.10	0.03	0.28	0.20	0.13
House sparrow	0.64	0.82	0.11	2.62	2.28	0.77
Teal	0.01	0.02	0.00	0.00	0.00	0.00
Wigeon	0.02	0.00	0.00	0.00	0.00	0.00
Sparrowhawk	0.00	0.00	0.01	0.00	0.00	0.01
Hobby	0.00	0.00	0.00	0.00	0.00	0.01
Woodcock	0.01	0.00	0.00	0.00	0.00	0.00
Green sandpiper	0.00	0.00	0.01	0.01	0.00	0.01
Redshank	0.00	0.00	0.00	0.00	0.01	0.00
Stock dove	0.00	0.01	0.00	0.02	0.02	0.03
Cuckoo	0.00	0.00	0.00	0.00	0.02	0.00
Tawny owl	0.01	0.00	0.00	0.00	0.00	0.00
Short-eared owl	0.01	0.00	0.00	0.00	0.00	0.00
Green woodpecker	0.00	0.00	0.00	0.00	0.00	0.00
Great spotted woodpecker	0.00	0.00	0.00	0.00	0.00	0.00
Jay	0.00	0.01	0.00	0.00	0.00	0.00
Mistle thrush	0.00	0.01	0.00	0.02	0.01	0.01
Redwing	0.03	0.32	0.00	0.00	0.00	0.00
Wheatear	0.00	0.00	0.00	0.00	0.01	0.01
Blackcap	0.00	0.00	0.00	0.01	0.00	0.01
Whitethroat	0.00	0.00	0.00	0.09	0.06	0.04
Lesser whitethroat	0.00	0.00	0.00	0.00	0.00	0.00
Willow warbler	0.00	0.00	0.00	0.05	0.06	0.03
Chiffchaff	0.00	0.00	0.00	0.00	0.00	0.00
Spotted flycatcher	0.00	0.00	0.00	0.01	0.02	0.01
Yellow wagtail	0.01	0.00	0.00	0.05	0.07	0.01
Tree sparrow	0.00	0.00	0.00	0.01	0.00	0.00
Kingfisher	0.00	0.00	0.00	0.00	0.00	0.00
Garden warbler	0.00	0.00	0.00	0.02	0.00	0.00
Whinchat	0.00	0.00	0.00	0.00	0.00	0.00

FARMLAND HEDGEROWS: HABITAT, CORRIDORS OR IRRELEVANT? A SMALL MAMMAL'S PERSPECTIVE

T. E. Tew
Wildlife Conservation Research Unit, Department of Zoology, Oxford University, Oxford, UK
Current address: *Vertebrate Ecology and Conservation Branch, Joint Nature Conservation Committee, Monkstone House, City Road, Peterborough, PE1 1JY, UK*

Introduction

Although Britain's native small mammal species evolved in deciduous woodland, all have, to some extent, been successful in colonizing the agricultural ecosystem. For some species, the presence of hedgerows is an essential prerequisite, whilst others appear able to exploit the open arable landscape. This paper will first consider the habitat requirements of all the common small mammal species, before presenting new data on the most widespread arable small mammals - the bank vole and wood mouse.

The data presented are the result of a three year study into the ecology of small mammals and their predators conducted mainly on arable farmland at the Oxford University Farm, Wytham, Oxfordshire. The study site is shown in Figure 8.1, where solid lines show the configuration of hedgerows and the hatched area shows woodland. Over two of these fields, sown with wheat and barley, respectively, an extensive small mammal live-trapping grid (292 traps spaced at 24 metre intervals covering 15 hectares) was implemented for four nights each month every month. These two fields are shown in full outline in Figures 8.1, 8.3, 8.5, 8.8, etc. Live-trapping was similarly conducted at another arable study site (Newbury Field), which was similar to Wytham in every respect except for its isolation (>500 metres) from woodland. Live-trapping and radio-tracking methodology can be found in detail elsewhere (Tew and Macdonald, 1993).

Shrews

Three species of shrew may be found in the hedgerows of the arable ecosystem, the common shrew (*Sorex araneus*), pygmy shrew *(Sorex minutus)* and water shrew *(Neomys fodiens)*. Both common and pygmy shrews are widespread wherever there is ground cover and are frequently live-trapped in hedgerows. As cover in the cereal fields increases from May onwards, common shrews are also occasionally caught away from the hedgerow, although generally within 20 metres of it (Figure 8.2). Most of these captures will be of animals nesting in the hedgerow but foraging in the field, whilst a few may be shrews dispersing between suitable habitats. Shrews are opportunistic feeders, taking a wide variety of invertebrate prey (Churchfield, 1991a) in direct relation to prey abundance (Churchfield, 1982), and many such invertebrates are seasonally available in the lowland cereal fields of southern England (Aebischer, 1991).   The restricted distribution of shrews away from hedgerows in the summer may be limited by their poor burrowing abilities (Churchfield, 1991*a*) and the relative paucity of other small mammal burrows suitable for colonization.

Figure 8.1   Capture locations of common shrews (*Sorex araneus*) in hedgerows and fields at the main study site at Wytham, for winter and summer periods

The field to the left was sown to winter wheat, that to the right to spring barley. Live traps were set in a regular grid pattern at 24 metre intervals throughout the fields, and at 24 metre intervals in the hedgerows.

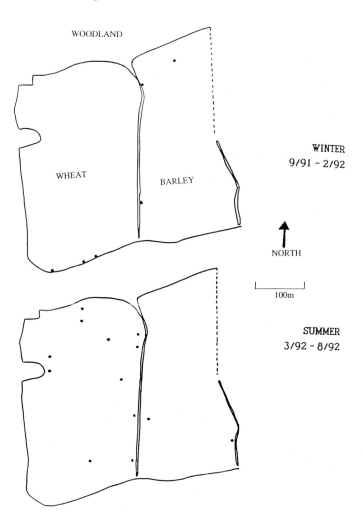

Water shrews are also occasionally caught on arable land, at great distances (>1 kilometre) from the streams and ponds that are their preferred habitat (Churchfield, 1991*b*), but only ever in ditches and hedgerows, and individuals are never captured more than one or two days consecutively. It is likely, therefore, that these captures are of dispersing animals and that hedgerows and ditches are important corridors between suitable habitats for these animals.

Figure 8.2   Numbers of bank voles (*Clethrionomys glareolus*) captured in the hedgerows and fields at Wytham (details as for Figure 8.1)

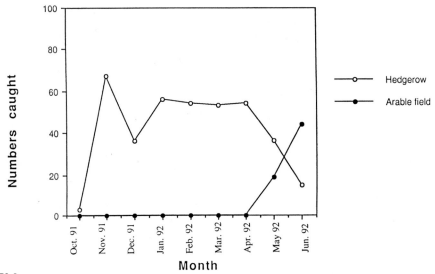

Voles

Two species of vole are caught on farmland, the field (or short-tailed) vole (*Microtus agrestis*) and bank vole *(Clethrionomys glareolus)*. As with the shrews, ground cover is an important habitat requirement (Southern and Lowe, 1968; Gurnell, 1985) and both species are restricted in their distribution on arable land. The field vole prefers rough, ungrazed pasture; it may occur in the grassy banks of arable hedgerows, but only at relatively low densities and in this study has never been caught away from the hedgerow in cereal fields.

The bank vole's preferred habitat is deciduous woodland but, like the common shrew, it is commonly caught in arable hedgerows throughout the year (Figure 8.3). Previous trapping studies on arable land have suggested that the bank vole is never caught away from hedgerows (Pollard and Relton, 1970; Jeffries *et al.*, 1973; Loman, 1991; Greig-Smith, 1991), but in this study, as the cover afforded by the crop increased throughout the summer, the bank voles frequently foraged into the field (Figure 8.3), although rarely more than 25 metres from a hedgerow (Figure 8.4). Food availability is likely to be low in the crop, since the preferred foods of bank voles are fleshy fruits and seeds and the leaves of woody plants (Watts, 1968; Hansson, 1985), but it is possible that the invertebrate fauna of the cereal field represents an alternative to the pre-fruiting hedgerow.

Figure 8.3   **Capture locations of bank voles (*Clethrionomys glareolus*) in hedgerows and fields for winter and summer periods (details as for Figure 8.1)**

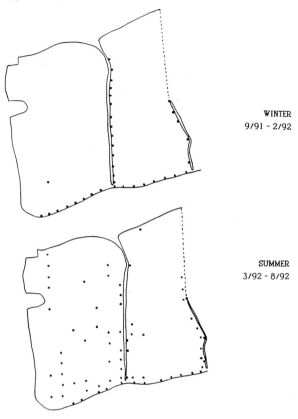

WINTER
9/91 - 2/92

SUMMER
3/92 - 8/92

Field observations indicate that bank voles in cereal fields may be directly vulnerable to the combine harvester, as well as to the indirect effects of harvesting (see below for wood mice). Similarly, observations of bank vole deaths have been made during the post-harvest burning of stubble. Direct observations suggest that their relative lack of agility and the shallowness of their burrow system increase the vulnerability of bank voles to the combine harvester and stubble fire, respectively. In both these cases, the bank voles appear to fare very much worse than the wood mice (Tew and Macdonald, 1993).

Mice

Three species of mice may be found in the hedgerows, the house mouse (*Mus domesticus*), harvest mouse (*Micromys minutus*), and wood mouse (*Apodemus sylvaticus*).

Figure 8.4   **Numbers of wood mice (*Apodemus sylvaticus*) captured at two arable sites; one adjacent to woodland (Wytham - Figure 8.1), the other >500 metres from woodland (Newbury); trapping methodologies were identical at both sites**

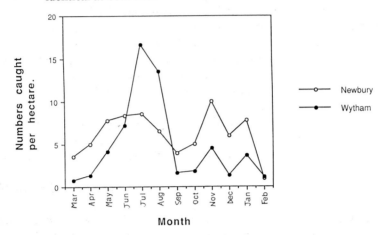

The house mouse is essentially commensal with man and competes poorly with other small mammal species away from human habitation (Dueser and Porter, 1986; Berry, 1991). It was once the third commonest small mammal on arable land, behind the wood mouse and bank vole (Southern and Laurie, 1946), but its numbers have decreased rapidly with the disappearance of corn ricks brought about by modern combining technology. Although still apparently common on agricultural land in N.W. Scotland (Delany, 1961), in southern England it is rarely caught on farmland away from farm buildings. Occasionally, however, juveniles are live-trapped moving along hedgerows in the autumn and early winter. These are probably animals dispersing from high-density populations, with the hedgerows, once again, serving as corridors.

The harvest mouse favours tall, dense vegetation (Harris, 1979), and cereal crops are a good habitat throughout the summer, although direct measures of abundance are problematical to collect, since harvest mice live in the stalk zone of the cereal and are rarely live-trapped on the ground. Although its historical distribution is uncertain (Harris, 1979; Harris and Trout, 1991), the harvest mouse is another rodent that has been badly affected by changes in modern agricultural practice. The rapidity of modern harvesting causes direct mortality and the absence of unthreshed corn ricks, in which harvest mice used to overwinter in large numbers (Rowe and Taylor, 1964), is highly disadvantageous. Because of this, hedgerows are important winter refuges.

The wood mouse is the most adaptable of all the small mammal species, and its catholic dietary (Hansson, 1985, for review) and habitat (Flowerdew, 1985, for review) preferences have allowed it successfully to colonize the entire arable ecosystem. In modern Britain, wood mice are true cereal-dwellers and are caught on open arable land great distances away from hedgerow or woodland at all times of the year (Pollard and Relton, 1970; Jeffries *et al.*, 1973; Green, 1979; Tew, 1989; Loman, 1991; Greig-Smith, 1991).

Arable wood mice

There may be three distinct wood mouse population types in the arable ecosystem. Woodland blocks on farmland certainly support high densities of wood mice (as well as bank voles) throughout the year, with population dynamics (Kikkawa, 1964) similar to those of deciduous woodland (Flowerdew, 1985). It is likely that 'surplus' animals from this habitat disperse into adjoining cereal fields and constitute part of the second population type - that of arable land near to woodland (Kikkawa, 1964). The third population type is a purely arable one that is self-sustaining and does not require seasonal immigration from other habitats to maintain numbers (Green, 1979; Tew, 1989; Loman, 1991).

The numbers of animals live-trapped at two different arable study sites is shown in Figure 8.5. One population (Wytham) was adjacent to woodland, whilst the other (Newbury) was >500 metres from the nearest woodland and separated from it by a road.

Figure 8.5   Spring population dynamics of wood mice in three different habitats

Open circles refer to the 'Newbury' (isolated arable) population; whilst closed circles and square symbols refer to a two other trapping grids, on the same farm as 'Newbury', on arable land adjacent to woodland and actually within deciduous woodland, respectively. Trapping methodologies were similar throughout.

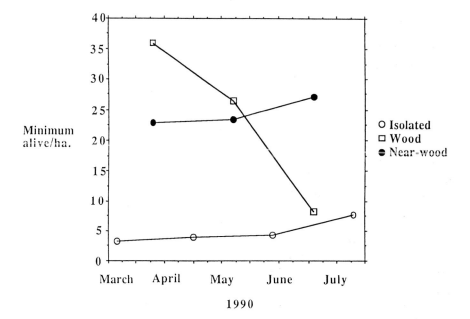

Overwintering densities were actually higher on the arable land furthest from the woodland, probably due to the lack of a more suitable habitat nearby. Populations on the more isolated arable land rise steadily until late summer as young of the year join the population. By contrast, mouse populations on arable land that is near woodland rise extremely rapidly in early summer as animals migrate out of the woodland (Figure 8.5), with the reverse effect on the adjacent woodland population (Figure 8.6). In 'typical' woodland habitat, there is often a drop in survival associated with increased aggression at the onset of the breeding season (Flowerdew, 1985; Montgomery and Gurnell, 1985) and the presence of nearby arable land to accommodate the dispersion of animals may reduce mortality.

Figure 8.6   Capture locations of wood mice in hedgerows and fields at three different times of the year (details as for Figure 8.1)

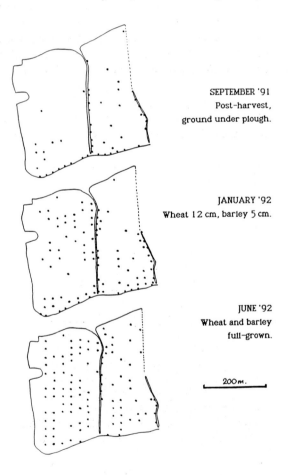

SEPTEMBER '91
Post-harvest,
ground under plough.

JANUARY '92
Wheat 12 cm, barley 5 cm.

JUNE '92
Wheat and barley
full-grown.

200 m.

The autumnal harvest of the cereal crop is a traumatic period for the arable wood mice and post-harvest mortality is high, caused mainly by post-harvest predation (Tew and Macdonald, 1993). Predation rises as a consequence of the removal of the cover afforded by the crop and the susceptibility to both aerial (tawny owl, *Strix aluco*) and terrestrial (weasel, *Mustela nivalis*) predators. As the cereal fields become dramatically sub-optimal after harvest, many mice reverse their migration back into nearby woodland (Kikkawa, 1964). Once again, it is unclear whether such moves are resisted by the woodland residents and, in any case, the number that survive to make this move after the effects of harvest-related mortality may be relatively small (Tew and Macdonald, 1993).

Whether through emigration or predation arable mouse numbers drop markedly at harvest (Greig-Smith, 1991; Tew and Macdonald, 1993) producing population dynamics different to those of woodland, in which animal numbers continue to rise into early winter (Flowerdew, 1985). Clearly, however, not all mice leave the arable fields during the winter and a proportion overwinter on arable land, even where woodland is proximate. In this study, animals continued to be trapped on the open fields (Figure 8.7) throughout the year, in accordance with previous work (Pollard and Relton, 1970; Green, 1979; Greig-Smith, 1991; Loman, 1991), and it seems likely that there is sufficient food available (waste grain, invertebrates) away from the hedgerows (Green, 1979).

Figure 8.7 (a)   **Percentage of wood mice captured in the cereal fields that were also caught in the hedgerows (details as for Figure 8.1)**

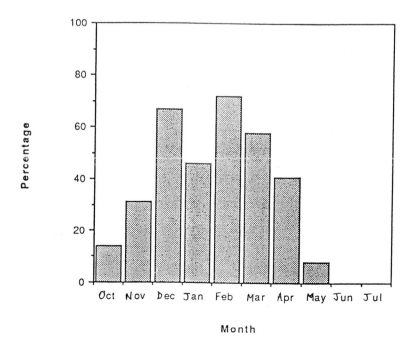

Month

Figure 8.7 (b)   **Percentage of wood mice captured in the hedgerows that were also caught in the cereal field.**

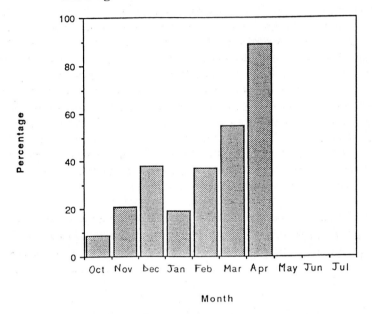

Month

Figure 8.7 (c)   **Numbers of wood mice captured in hedgerows and field.**

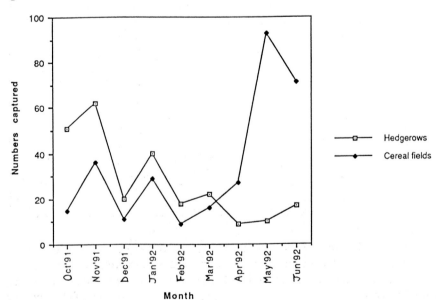

As with other species, hedgerows are an important part of the arable habitat for the wood mice throughout the winter. Between October and April up to 70 per cent of the mice caught on the open field were also caught in the hedgerow (Figure 8.7a). Between October and February only 30 per cent of the mice caught in the hedgerow were also using the field (Figure 8.7b); this percentage then rose rapidly in March and April as animals moved out of the hedgerow into the cereal crop. Thus, throughout the winter it appears that a large section of the population are able to use the hedgerow only, without recourse to the field, whilst fewer are able to use the field without recourse to the hedgerow. At this time of year, the fruits, berries and permanent invertebrate supply of the hedgerows are likely to be an increasingly important food source for the arable mice.

From May onwards there is no interchange between the hedgerow and field and the two populations are discrete (Figure 8.7a, b). Clearly, at this time of year the cereal fields can provide all the resources necessary and the field population of mice, in contrast to that of the hedgerow, increases markedly (Figure 8.7c).

Figure 8.8   Radio-locations and home-range outlines for a male wood mouse during both winter (squares) and spring (circles) (hedgerows denoted by bold lines, same study site as Figure 8.1, axes scale in metres)

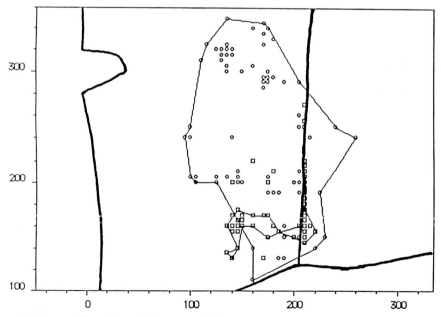

Further confirmation of these population processes are provided by the study of movement patterns of individual mice. An example is given (Figure 8.8) of a male mouse radio-tracked for five nights both in February and April. The mouse lived largely in the hedgerows over the winter, with occasional forays to the open field, and then expanded his range and moved out into the cereal field in the spring as the crop grew. This pattern is generally true for all the arable mice, and examples are given of typical home-range distributions for both winter and summer (Figures 8.9, 8.10).

Figure 8.9   **Radio-locations and home-range outlines for four wood mice during winter (details as Figure 8.8)**

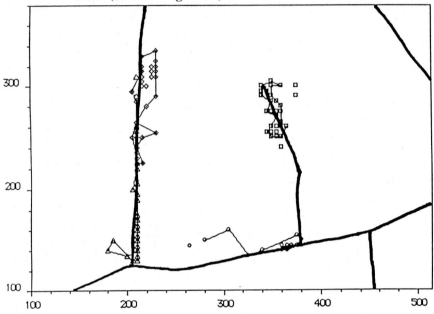

Figure 8.10   **Home-range outlines for four wood mice during summer (details as Figure 8.8)**

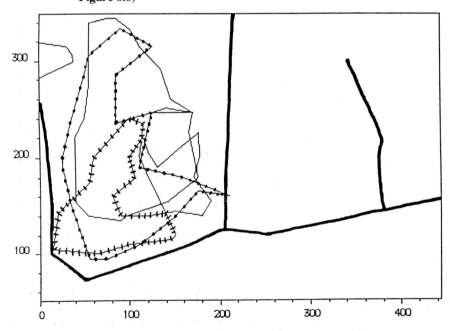

The social system of arable mice during the summer is one of female defence polygyny (Tew and Macdonald, 1994), with females defending small (0.5 hectare) territories intra-sexually and males occupying large (1.5 hectare) home-ranges that overlap both inter- and intra-sexually. Seasonal territoriality by females may explain the seasonal nature of the interchange between the hedgerow and field populations.

Throughout the summer months, the hedgerows may in fact hold greater predation pressure for the mice than the arable fields. The major predator of small mammals on farmland is the weasel, which utilizes farmland hedgerows highly preferentially (Figure 8.11). Since both shrews and voles are largely restricted to hedgerows, the overall density of small mammals is much higher there than in the field and so this is not surprising. Weasels on farmland are essentially diurnal burrow hunters so that mice nesting in hedgerows would appear to be at considerably greater risk than those that nest in the field. Similarly, both diurnal and nocturnal avian predators also focus on the hedgerow (direct observation).

Figure 8.11   **Radio-locations of two male weasels (*Mustela nivalis*), over farmland at Wytham. Hedgerows denoted by solid lines, woodland by cross-hatching. The range of one male (open squares) encompassed the main small mammal live-trapping site (approximate coordinates 4700087000). The axes scales are in metres**

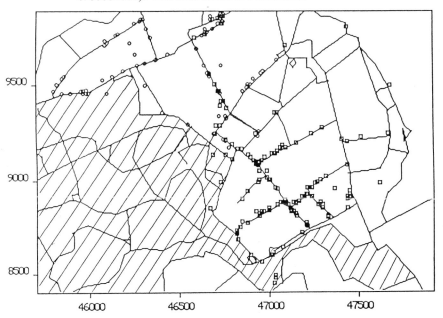

Hedgerow presence and quality

Few studies have been conducted on the effects of hedgerow quality on small mammal abundance. Boone and Tinklin (1988) found higher densities of both wood mice and bank voles in hedges with thick understoreys than in those with thin understoreys, although trapping was only conducted for five months and sample sizes were small. Plesner Jensen (personal communication) studied the effects of physical features (hedges, ditches, tracks, etc.) on the number of small mammals captured on experimental field margins (Smith and Macdonald, 1989) and found predictable inter-specific differences. The number of wood mice captured on the margins was largely unaffected by the surrounding physical features except after harvest, when the presence of a hedgerow correlated with greater abundance. Bank voles, on the other hand, were affected to a much greater extent, with year-round positive correlations of animal numbers to hedgerow presence. Similarly, both common, pygmy and water shrew numbers were all positively correlated with hedgerow presence to a greater or lesser extent.

Conclusions

All the available data on small mammals suggests that hedgerows are an important feature of the arable landscape, although their function and the extent of their importance varies between species.

For shrews and voles arable hedgerows serve as permanent habitats, although the cereal fields are exploited for foraging excursions when cover is available. Hedgerows must also be important dispersal corridors for such species, particularly the water shrew, in a similar way that fence-rows relieve the barrier created by open fields around woodland blocks for the North American white-footed mouse, *Peromyscus leucopus* (Wegner and Merriam, 1979).

Harvest mice are similarly dependant on hedgerows, although only seasonally, serving as important winter refuges.

Wood mice are least affected by hedgerow presence. Wood mice are an opportunistic and highly motile species that have adapted well to the agricultural ecosystem and are well able to sustain populations without either hedgerows or woodland. Nevertheless, hedgerows are exploited by wood mice, particularly after harvest and during the winter, and their presence almost certainly allows higher population densities on farmland.

With such a concentration of small mammals in hedgerows it is not surprising that specialist predators, particularly the weasel, concentrate their hunting in this part of the habitat. Indeed, it is extremely unlikely that farmland would be able to support viable weasel populations without the presence of hedgerows.

Acknowledgements

I would like to thank all the project staff and fieldworkers who have helped in the collection of the live-trapping and radio-tracking data, particularly Dr Ian Todd, Liz Brown, Dr Mick Pullen, Justine Freeth, Lauren Blencowe, Sue Waugh, Cindi Birkle, and Anna Povey. The work was funded by the Joint Agricultural and Environmental Programme (JAEP), grant reference number GST/02/479.

References

Aebischer NJ. (1991) Twenty years of monitoring invertebrates and weeds in cereal fields in Sussex. In: Firbank LG, Carter N, Darbyshire JF, Potts GR, eds. *The ecology of temperate cereal fields.* Oxford: Blackwell Scientific Publications, 305-32.

Berry RJ. (1991) The house mouse. In: Corbett GB, Harris S, eds. *Handbook of British mammals.* Oxford: Blackwell Scientific Publications, 239-47.

Boone GC, Tinklin R. (1988) Importance of hedgerow structure in determining the occurrence and density of small mammals. *Aspects of Applied Biology* **16,** 73-8.

Churchfield S. (1982) Food availability and the diet of the common shrew, *Sorex araneus,* in Britain. *Journal of Animal Ecology* **51,** 15-28.

Churchfield S. (1991*a*) The common shrew. In: Corbett GB, Harris S, eds. *Handbook of British mammals.* Oxford: Blackwell Scientific Publications, 51-8.

Churchfield S. (1991*b*) The water shrew. In: Corbett GB, Harris S, eds. *Handbook of British mammals.* Oxford: Blackwell Scientific Publications, 64-8.

Delany MJ. (1961) The ecological distribution of small mammals in north-west Scotland. *Proceedings of the Zoological Society, London* **137,** 107-26.

Dueser RD, Porter JH. (1986) Habitat use by insular small mammals: relative effects of competition and habitat structure. *Ecology* **67,** 195-201.

Flowerdew JR. (1985) The population dynamics of wood mice and yellow-necked mice. *Symposium of the Zoological Society of London* **55,** 315-38.

Green RE. (1979) The ecology of wood mice (*Apodemus sylvaticus*) on arable farmland. *Journal of Zoology, London* **188,** 357-77.

Greig-Smith PW. (1991) The Boxworth experience: effects of pesticides on the fauna and flora of cereal fields. In: Firbank LG, Carter N, Darbyshire JF, Potts GR, eds. *The ecology of temperate cereal fields.* Oxford: Blackwell Scientific Publications, 333-71.

Gurnell J. (1985) Woodland rodent communities. *Symposium of the Zoological Society of London* **55,** 377-411.

Hansson L. (1985) The food of bank voles, wood mice and yellow-necked mice. *Symposium of the Zoological Society of London* **55,** 141-68.

Harris S. (1979) History, distribution, status and habitat requirements of the harvest mouse (*Micromys minutus*) in Britain. *Mammal Review* **9,** 159-71.

Harris S, Trout RC. (1991) The harvest mouse. In: Corbett GB, Harris S, eds. *Handbook of British mammals.* Oxford: Blackwell Scientific Publications, 233-9.

Jeffries DJ, Stainsby B, French MC. (1973) The ecology of small mammals in arable fields drilled with winter wheat and the increase in their dieldrin and mercury residues. *Journal of Zoology, London* **171,** 513-39.

Kikkawa J. (1964) Movement, activity and distribution of the small rodents *Clethrionomys glareolus* and *Apodemus sylvaticus* in woodland. *Journal of Animal Ecology* **33,** 259-99.

Loman J. (1991) The small mammal fauna in an agricultural landscape in southern Sweden, with special reference to the wood mouse *Apodemus sylvaticus. Mammalia* **55,** 91-6.

Montgomery WI, Gurnell J. (1985) The behaviour of *Apodemus. Symposium of the Zoological Society of London* **55,** 89-115.

Pollard E, Relton J. (1970) A study of small mammals in hedges and cultivated fields. *Journal of Applied Ecology* **7,** 549-57.

Rowe FB, Taylor EJ. (1964) The numbers of harvest mice (*Micromys minutus*) in corn ricks. *Proceedings of the Zoological Society, London* **142,** 181-5.

Smith H, Macdonald DW. (1989) Secondary succession on extended arable fields: its manipulation for wildlife benefits and weed control. *Brighton Crop Protection Conference - 1989 - Weeds.* Farnham: BCPC Publications, 1063-8.

Southern HN, Laurie EMO. (1946) The house mouse (*Mus musculus*) in corn ricks. *Journal of Animal Ecology* **15,** 135-49.

Southern HN, Lowe VPW. (1968) The pattern and distribution of prey and predation in tawny owl territories. *Journal of Animal Ecology* **37,** 75-97.

Tew TE. (1989) The behavioural ecology of the wood mouse *(Apodemus sylvaticus)* in the cereal field ecosystem. Unpublished DPhil Thesis, University of Oxford.

Tew TE. (1992) Radio-tracking arable wood mice. In: Priede IG, Swift SM, eds. *Wildlife telemetry.* London: Ellis Horwood, 532-9.

Tew TE, Macdonald DW. (1993) The effects of harvest on arable wood mice (*Apodemus sylvaticus*). *Biological Conservation* **65,** 279-83.

Tew TE, Macdonald DW. (1994) Dynamics of space use and male vigour amongst wood mice, *Apodemus sylvaticus,* in the cereal ecosystem. *Behavioural Ecology and Sociobiology* **34,** 337-45.

Watts CHS. (1968) The foods eaten by wood mice (*Apodemus sylvaticus*) and bank voles (*Clethrionomys glareolus*) in Wytham woods, Berkshire. *Journal of Animal Ecology* **37,** 25-41.

Wegner JF, Merriam G. (1979) Movement of birds and small mammals between a wood and adjoining farmland habitats. *Journal of Applied Ecology* **16,** 349-55.

CHAPTER 9

FLORISTIC DIVERSITY, MANAGEMENT AND ASSOCIATED LAND USE IN BRITISH HEDGEROWS

R. P. Cummins and D. D. French
Institute of Terrestrial Ecology, Hill of Brathens, Banchory, Kincardineshire AB31 4BY, Scotland, UK

Introduction

Although there have been many studies of the floristics of hedgerows, they have generally been of small scale (i.e. at the county level or smaller) and have tended to concentrate on the hedges themselves at the expense of the herbaceous species of the hedge-bottom which can be just as important ecologically. The studies also vary widely in their methodologies which severely limits the scope for integrating the various databases.

Consequently, there has previously been no national overview of hedgerow floristics and their relationships with management and land-use. In particular, there have been no objective classifications of hedges/hedge-bottoms that can be applied throughout Great Britain to provide a common basis for discussion and research. Without such objective classifications, it is difficult for policy-makers or conservationists to make decisions in a national context.

This paper describes some of the data on hedgerows collected during field surveys of the countryside in general, throughout Great Britain, carried out by the Institute of Terrestrial Ecology (ITE) in 1978 and 1990. The surveys were not specifically of hedgerows, but do provide the only nation-wide database of hedgerows currently available. Consequently, the data were of direct relevance to executive bodies such as the Department of the Environment (DoE) and the Countryside Commission in the development of policies for the protection and maintenance of hedgerows, such as the Hedgerow Incentive Scheme. The results presented here are extracted from a report subsequently commissioned by DoE (Cummins *et al.*, 1992) and focus on hedgerow classification and land-use.

Throughout we will use 'hedgerow' as a generic term for the feature as a whole; 'hedge' will refer to the woody species component and 'hedge-bottom' to the herbaceous component at the foot of the hedge.

Methods

The sampling strategy was the same in both years, namely, a random sample of Ordnance Survey 1 kilometre squares throughout GB, stratified according to the 32 classes of the ITE Land Classification (Bunce *et al.*, 1983). In 1978, 256 squares were sampled and this was increased to 508 squares in the 1990 survey which was funded by

DoE and the Natural Environment Research Council with support from the Nature
Conservancy Council. The distribution of the survey squares and of those containing
hedges is shown in Figure 9.1.

**Figure 9.1   Distribution of survey squares in 1990 showing those that contained
hedge plots**

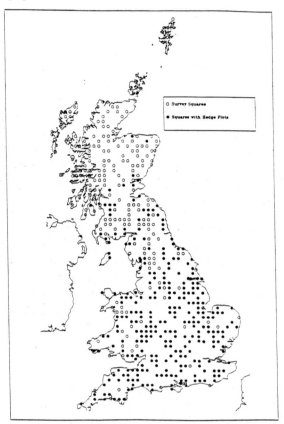

    In each square, all land cover and landscape features were mapped, including all
hedgerows except those forming part of a domestic curtilage. Floristic data were obtained
from 10 metre x 1 metre quadrats which were set out with the long side running along
the centre line of the hedge. Thus only one side of the hedge was sampled. All species
that were rooted in the quadrat were recorded, along with their percentage cover (where
greater than 5%). Plots were relocated if there was not a clear metre between the centre
of the hedge and another linear feature, e.g. a ditch or road. This ensured that the flora
sampled was only related to hedgerows, and not from a mixture of very different
habitats. In each square, two hedge plots were established using a set procedure to ensure
that the plots were well separated within the square. In 1978, 278 quadrats were
recorded and their locations marked on 1:10,000 maps. Of these, 259 were reliably
located (from their mapped positions and measured distances on the ground) and re-

recorded in 1990. The same procedure was used to establish plots in squares which had not been surveyed in 1978. Also in 1990, up to 5 more quadrats per square were recorded alongside boundaries in general. Sometimes these boundaries were formed by hedgerows and such data (a further 659 quadrats) have been included in the analyses.

Analyses

Classification

Classification of the hedgerows used the statistical   procedure TWINSPAN (Hill, 1979*a*). All 1176 plots from 1978 and 1990 were lumped together for analysis, thus ensuring that any hedgerow types which were present in only one year were not omitted. Initial TWINSPAN analyses of all the species data together (woody and herbaceous) gave a classification which was confused by the large amount of information present. To get interpretable results, the data on hedges were analysed separately from those on hedge-bottoms. Hedges were classified using percentage cover of the woody species (i.e. including roses but not soft-wooded species such as bramble). Due to the large number of herbaceous species recorded, it was possible to classify the hedge-bottoms using only presence/absence data.

It is important to note that the classifications operate on the *balance* between different species and/or their cover values. The presence of a new species, or a change in the cover of an existing species, may alter the classification of a hedge/hedge-bottom even though its overall appearance is not markedly changed. Hence a species which typifies a class may not necessarily be dominant, i.e. have the greatest cover in a stretch of hedgerow. Consequently, we have distinguished between 'dominant', 'predominant' (i.e. the indicator species is very frequent but not necessarily at a high cover value) and 'present' (just the presence of the species was sufficient to classify a hedge).

Associations of species

The classification keys use only a small proportion of all the species analysed (the key to hedge classes, for example, uses only 20 of the 55 available species, and never more than 7 species for any one division) and give little information about the types of vegetation in the different classes. Floristic descriptions of the classes could be very complex, especially when there are over 200 herbaceous species (some of which are aggregates). Here the descriptions have been simplified by statistically grouping closely associated species together into a relatively small number of 'clusters' and then examining the occurrence of the various clusters in the different classes of hedges/hedge-bottoms.

The species clusters were obtained by inputting scores from a DECORANA analysis (Hill, 1979) into Nearest Neighbour analysis (woody species) or Ward's minimal variance analysis (herbaceous species) using the SAS CLUSTER procedure (SAS Institute, 1989).

Results

Classification of hedges

Initially, the classification split off a series of less common hedges - classes H 1, 2, 3, 7, 8 - all but one having fewer than 20 plots in the sample (Table 9.1). The amount of hawthorn present was then a strong factor in separating the principal hedge types (H 4, 5, 6) and in subdividing the biggest class (H 4). Eleven classes of hedges resulted.

Table 9.1   TWINSPAN hedge classes, types and percentage frequency

Class	Type	Percentage of plots
H 1	Mostly planted non-native species	1
H 2	Wild privet present	0.4
H 3	Beech dominant	2
H 4a	Hawthorn dominant	47
H 4b	Mixed hawthorn	5
H 4c	Elder/hawthorn	3
H 5a	Willow or rose dominant	0.5
H 5b	Mixed hazel predominant	13
H 6	Blackthorn predominant	23
H 7	Elm predominant	4
H 8	Gorse dominant	0.7

Classification of hedge-bottoms

Level 3 of the TWINSPAN hierarchy gave four acceptable composite classes (HBs) which we consider to be adequate for present purposes. Hegarty and Cooper (1994) found very similar classes from their work in Northern Ireland.

Not surprisingly, several of the very common species occur in two or more adjacent HBs, albeit at different frequencies and percentage cover values: at least two-thirds of the species prevalent in groups 1 and 2 were also prevalent in other groups. In contrast, only one-third or less of the common species in HBs 3 and 4 were also frequently present in HBs 1 and 2. Despite this overlap, a DECORANA analysis suggested that the herbaceous species associations are typical of the following types of land-use/habitats:

- HB 1   arable cropland;
- HB 2   other intensively managed ground (mainly lowland; possibly rotational);
- HB 3   less intensively managed grasslands and rough grazings;
- HB 4   woodland vegetation.

The relationships between the occurrence of these HBs and the adjacent land-use will be discussed later.

Both the classifications outlined above (along with their associated keys) are statistical divisions of species associations which form a series of continua ranging from various mono-specific types to very mixed vegetation. As with any continua, subdivision will result in the extremes of one division being very similar to those of adjacent classes. Therefore, it must be stressed that the descriptive terms given to the different classes are not to be used casually. The different classes have been precisely defined and the appropriate keys must be used to avoid mis-classification.

Floristics of the hedgerow classes

The overall floristics of each class is indicated by their association with different species clusters. The strength of the association is determined by the degree to which each cluster of species occurs more or less frequently than one would expect statistically for that class.

Woody species: Clustering of woody species over all the hedge classes showed two important groupings:

- Cluster 1: Field maple, sycamore, alder, hazel, hawthorn, spindle, ash, holly, birch;
- Cluster 2: Blackthorn, roses, elder, broom, guelder rose, oak.

The first cluster contains essentially long-lived, woodland species that are all native, except for sycamore which was introduced in the fourteenth or fifteenth century. The species in cluster 2 are also native but, with the exception of oak, do not live as long as those in cluster 1 and are associated with more open scrubby habitats. Other clusters tended to be of incidentals associated with one strongly dominant species, e.g. beech.

Although mixed-hawthorn and mixed-hazel hedges have several species from both clusters, they are the only classes with a wide representation of the woodland species in cluster 1. They are therefore of ecological importance, not only for the type of woody species present, but also for their diversity which, respectively, averaged 4.4 and 3.6 woody species per 10 metre plot. This compares with 2.0 and 3.2 species for the two most common hedge classes - hawthorn-dominant and blackthorn.

Herbaceous species: Twelve herbaceous species clusters were identified and can be summarized by the habitat-type which their component species have in common (Table 9.2).

Despite the physical dominance of the woody components of hedges, Table 9.2 clearly shows that the vegetation in hedge-bottoms is not confined to associations typical of woodlands. In fact the occurrence of species from the woodland clusters (4b and 6) together accounted for only 15 per cent of the total. A greater proportion (25-35 per cent) contained species typical of abandoned or derelict ground (clusters 5 and 8b) or grassland (clusters 1a, 1b, 2, and 4a). The associations between the above clusters and the different types of hedge-bottom are shown in Table 9.3.

The detailed results in Table 9.3 support the more general conclusions obtained from the DECORANA interpretation.

Table 9.2   The occurrence of herbaceous species in different clusters as a percentage of all herbaceous species records

Cluster	Habitat type	Typical species	% of records
	Meadow	*Dactylis glomerata*	19
		Cirsium arvense	
		Festuca rubra	
1b	Short-term grassland	*Lolium perenne*	8
		Poa trivialis	
		Stellaria media	
2	Long-term/established pasture	*Ranunculus repens*	5
		Trifolium repens	
		Veronica chamaedrys	
3	Nitrogen/phosphorus-rich areas	*Lamium album*	1
		Conium maculatum	
		Vicia sativa	
4a	Acid grassland	*Agrostis capillaris*	3
		Anthoxanthum odoratum	
		Chamaenerion angustifolium	
4b	Acid woodland	*Holcus mollis*	5
		Pteridium aquilinum	
		Digitalis purpurea	
5	Abandoned/derelict areas	*Galium aparine*	2
		Arrhenatherum elatius	
		Heracleum sphondylium	
6	Lowland wood/scrub	*Hedera helix*	10
		Glechoma hederacea	
		Brachypodium sylvaticum	
7	Arable weeds	*Myosotis spp.*	2
		Chenopodium spp.	
		Lamium purpureum	
8a	Humus-rich basophiles	*Urtica dioica*	14
		Rubus fruticosus agg.	
		Stellaria holostea	
8b	Southern 'derelict' areas	*Tamus communis*	3
		Arum maculatum	
		Solanum spp.	
9	Common weeds	*Centaurea spp.*	7
		Epilobium montanum	
		Viola riviniana	

Table 9.3   The relative frequency (observed versus expected) of plots in different combinations of herbaceous species clusters/hedge-bottom classes

Species cluster	Habitat type	Hedge-bottom (HB) TWINSPAN classes			
		1 Intensive arable	2 Other lowland	3 Grassland	4 Woodland
1a	Meadow	-		+	
1b	Short-term grassland	-		+	-
2	Long-term/established pasture	- - -		++	
3	N/P rich areas	+		- -	- - -
4a	Acid grassland	- - -	-	++	++
4b	Acid woodland	- -			+++
5	Abandoned/derelict areas	+		- -	- - -
6	Lowland wood scrub				+
7	Arable weeds	+		-	- - -
8a	Humus-rich basophiles				
8b	Southern 'derelict' areas			-	
9	Bare ground/common weeds				

+/- indicates observed value is at least 1.5 times, ++/-- at least twice, and +++/--- at least 3 times greater or less than expected. The absence of a sign indicates an average degree of association. Expected values determined from χ^2 contingency table.

(a) The species clusters positively associated with HB 1 are typical of arable land (i.e. those of nitrogen/phosphorus-rich areas and arable weeds) and also with species common on land that was formerly managed, but is now abandoned/derelict. There were negative associations with all the grassland clusters, including short-term grassland, and also with acid woodland.

(b) HB 2 has an average level of association with all the clusters except one (a negative association with species typical of acid grassland). This lack of characteristic clusters make this type of hedge-bottom difficult to identify in the field without the classification key.

(c) The herbaceous components of HB 3 can be generally summed up as 'grasslands' (including hay and silage production); the positive associations are with all four grassland clusters and negative associations with the clusters characteristic of arable areas and unmanaged ground.

(d) HB 4 was the only class with positive affinities for woodland species (clusters 4b and 6); conversely, there were negative associations with clusters typical of high management levels. The positive and negative associations here also suggest vegetation types that are common amongst low-input grazings and marginal ground.

These results suggest that the presence of a hedge may not be the dominant factor determining the composition of the herbaceous communities beneath it. Indeed, other analyses (data not given here) show that, of the more common types of hedge, only mixed hazel hedges are positively associated with the woodland type of ground flora (clusters 4b and 6) that one might expect to find.

We also examined the relationship between the hedge-bottom communities and the methods of trimming a hedge, hedge height and 'gappiness'. None of the results were significant, but these preliminary analyses were done on all hedges together; further analyses are required for individual classes of hedge. However, experimental work at ITE Monkswood also indicates that hedge management has a minimal effect on the ground flora (M.D. Hooper, personal communication).

So far, we have seen that neither the type of hedge nor the method of its management seems to have much effect on the composition of the ground flora - so what is the major controlling factor? The inference from the results above is that the adjacent land-use could be a dominant factor.

Relationships with land use

In previous sections, we have examined the associations between hedgerow classes and different habitats/land uses as indicated by the herbaceous vegetation. In this section, the land-cover maps drawn during the survey (see Methods) are used to determine directly the land-use adjacent to the quadrats.

Land-uses were divided into the following ten categories. Those in brackets were not used in the analyses because the sample sizes were too small.

Cereal crops	Trees/shrubs
Non-cereal crops	Amenity ground
Intensively-managed grassland	Road/track
Other managed grassland	[Urban]
Unmanaged grass/herb	[Water]

Hedge classes and land-use

A cross-tabulation of land-use categories with hedge classes gives 64 possible combinations. Only seven of these combinations occurred in amounts that were statistically different (two at $P<0.1$, five at $P<0.05$) from expected (chi-square contingency). In practice, the relationship between land-use and hedge-type is confounded by geographically determined variations in their distributions and the results show no consistent patterns which would enable us to draw firm conclusions about the effects, if any, of land-use on the type of hedge present. However, it is notable that the cereal and non-cereal croplands were each positively associated with a hedge type that is species-poor (elder-hawthorn and hawthorn-dominant, respectively) and negatively associated with a richer type of hedge (mixed-hazel and blackthorn, respectively).

Hedge-bottoms and land-use

The associations between different types of hedge-bottoms and land-uses are shown in Table 9.4, where land-uses have been arranged in order of management intensity. Land-uses have been excluded if there were no significant differences between the observed and expected numbers of plots for any of the different types of hedge-bottom.

Table 9.4   The percentage of plots in each hedge-bottom (HB) TWINSPAN group that were adjacent to different land uses

Land-use	Hedge-bottom (HB) TWINSPAN groups			
	1 Intensive arable	2 Other lowland	3 Grassland	4 Woodland
Cereals	50^{++}	17^{-}	3.0^{-}	4.3^{-}
Non-cereal crops	16^{++}	3.6^{-}	0.0^{-}	2.1^{-}
Intensively managed grass	9.1^{-}	14	27^{++}	18
Other managed grassland	11^{-}	50^{++}	4^{++}	46^{++}
Road/track	6.0	7.6	6.9	14^{++}
Unmanaged grassland	2.7	0.5^{-}	4.0	5.3

Superscripts indicate frequencies that were more (+) or less (-) than expected (χ^2 analysis. +/- indicates $P < 0.1$, ++/-- $P < 0.05$). Figures without superscripts were not significantly different from expected values.

There is a strong diagonal gradient across Table 9.4, showing that the intensity of land management adjacent to the hedgerow decreases from HB 1 to HB 4. The positive and negative associations shown here reflect fairly closely the associations between HBs and the 'habitat-type' of the species clusters (Table 9.3). In other words, the types of plants found in the hedge-bottom are similar to those that you would expect to find in the adjacent land-use. The similarity is particularly strong for HBs 1,3 and 4. Thus, there is clear evidence that land-use could have a major effect on the vegetation in the hedge-bottom. However, this conclusion is inferential and the true relationships can only be determined by experimentation.

Land-use and species richness

For this paper, the simple measure of species richness that we have used is the mean number of species per 10 metre x 1 metre plot. Obviously the types of species present are also important (e.g. a numerically diverse hedgerow may contain a lot of 'weed' species that might be considered as 'undesirable') but lack of space prevents consideration of this more complex topic.

Hedges: There was a large range in the richness of woody species according to land-use: the highest value (3.9 for hedges alongside 'tree/shrub') was 77 per cent greater than the minimum (2.2 for hedges adjacent to unmanaged grassland). The value for hedges adjacent to trees/shrubs was significantly greater (ANOVA/LSD, $P<0.1$) than in any other land-use except amenity ground. Conversely, hedges alongside unmanaged grassland or either type of cropland were significantly poorer ($P<0.1$) in woody species than elsewhere. None of the differences between other pairs of land-uses are statistically significant.

Hedge-bottoms: The pattern of herbaceous species richness across land-uses differed from that of the woody species and had a proportionally smaller range, the maximum being about 50 per cent more than the minimum. The richest hedge-bottoms were adjacent to the intensively managed grassland (15.8 species per plot) and 'other managed grassland' (15.0 species per plot). The poorest were alongside non-cereal crops (10.5 species per plot). However, most of the differences are not statistically significant due to the considerable intrinsic variability (the least significant difference ($P<0.1$) between any pair of land-uses was 3.8 species).

The 'total' species richness (woody + herbaceous) parallels that of the herbaceous species. Although an analysis of variance over all land-uses is highly significant ($P<0.001$), the few significant differences in pairs of land-uses are confined to those at opposite ends of the range.

Discussion

Due to the 'snapshot' nature of surveys, the effects of land-use on diversity can only be inferred; direct cause/effect relationships cannot be determined. This is particularly so with the woody species and there are many possible explanations for the few differences that there are between land-uses. For example, hedges with few woody species are more prevalent amongst croplands than one would expect statistically but this could be because only monotypic hedges are planted there, or perhaps the hedges are younger there than elsewhere and have not yet been invaded by other species.

However, hedges are a long-lived feature and their composition of established woody species is unlikely to be related to factors which fluctuate on a relatively short time-scale, e.g. land-use changing from pasture to cereals. Furthermore, agricultural practices are usually concentrated on the ground vegetation and it is only accidents such as spray drift which are likely to affect the diversity of woody species (we have already shown that the hedge class is not strongly related to hedge management *per se*). On the other hand, tillage, grazing, and spraying into the hedge-bottom can kill off young tree seedlings which might otherwise be recruited into the hedge, either as new species or as regeneration of existing species. This would partly account for the hedges in croplands being poor in woody species and it can be ameliorated by leaving wider headlands.

Many plants form large stores of dormant seeds in the soil ('seed banks') which can survive for many years and only germinate when the ground is disturbed. These reserves are an important source of regeneration. Tree species generally do not have large seed banks (Grime *et al.*, 1988). Therefore some hedges, especially in arable areas, may depend on an input of seeds from other sources for regeneration/ recruitment (hence the richness of hedges adjacent to trees/shrubs).

It is not possible to quantify the effects of all the above factors on the species diversity in hedges but it is important to be aware of them when, for example, ageing hedges by species counts: a hedge with, say, three woody species in a long-established arable area may actually be older than a hedge containing four species that is next to a woodland. Mixed hazel hedges were particularly rich in woody species and are obvious targets for conservation, not only for their ecological value but also because, according to Hooper's Rule (Hooper, 1970), they probably include a high proportion of old hedges which can be of historical importance.

Compared with the woody species, more substantial inferences can be drawn about the effects of land-use on herbaceous species because a direct comparison can be made between the vegetation in the hedge-bottom and that of the adjacent land. We have shown that the species associations in most hedge-bottoms resembled those of the adjacent land-use rather than the woodland-type vegetation expected in an 'unaffected' hedgerow.

The effects of land-use can be ameliorated by introducing wide headlands which can then be subjected to different management regimes from the remainder of the land (Smith and Macdonald, 1992). This practice would be particularly valuable in croplands where hedge-bottoms have a markedly poor flora. However, if a more diverse ground flora is to be re-established within, say, 1 metre of the hedge then the headland should be considerably wider than this so that the 1 metre strip itself is effectively buffered from land-use practices (Davis *et al.*, 1990). Even then, the seed bank of herbaceous species may be so depleted in some arable areas that a non-weedy hedge-bottom flora may not develop without assistance, e.g. by seeding with appropriate wild species.

Our species richness data show that very intensive land management and no management at all are both deleterious to the number of herbaceous species in hedge-bottoms. Hence increasing the headlands in arable situations may have only a small effect on the number of species present unless the headland is managed by cutting or grazing.

When deciding on management of the hedge-bottoms, it is important to consider also the types of species to be conserved - for example, the fairly high levels of grazing/cutting that would be needed to maintain a community of pasture species would be deleterious to a woodland-type community. The adverse effects of high levels of fertilizer should not be overlooked; they can promote the growth of many 'weed' species which will then out-compete the semi-natural vegetation types characteristic of some hedgerows.

Clearly the woodland-type species in HB 4 are predominantly associated with land uses that have low levels of management and which tend to be undisturbed over relatively long periods. The likelihood of road/tracksides being refuges is noteworthy.

References

Bunce RGH, Barr CJ, Whittaker HA. (1983) A stratification system for ecological sampling. In: Fuller RM, ed. *Ecological mapping from ground, air and space.* ITE symposium no.**10.** Cambridge: Institute of Terrestrial Ecology, 39-46.

Cummins RP, French DD, Bunce RGH, Howard DC, Barr CJ. (1992) *Diversity in British hedgerows*. Contract report to the Department of the Environment. Banchory: Institute of Terrestrial Ecology.

Davis BNK, Yates TJ, Lakhani KH. (1991) Spray drift impacts on non-target organisms. In: *Institute of Terrestrial Ecology Annual Report 1989-1990*. Huntingdon: Institute of Terrestrial Ecology, 40-3.

Grime JP, Hodgson JG, Hunt R. (1988) *Comparative plant ecology: a functional approach to common British species*. London: Unwin Hyman.

Hegarty, CA, Cooper A. (1994) The composition, structure and management of hedges in Northern Ireland. In: Watt TA, Buckley GP, eds. *Hedgerow management and nature conservation*. Wye College Press, Wye College, University of London, 153-4.

Hill MO. (1979*a*) *TWINSPAN - a FORTRAN program for arranging multivariate data in an ordered two-way table by classification of the individuals and attributes.*: Section of Ecology and Systematics, Cornell University, Ithaca, New York.

Hill MO. (1979*b*) *DECORANA - a FORTRAN program for detrended correspondence analysis and reciprocal averaging*. Ithaca, NY: Section of Ecology and Systematics, Cornell University, Ithaca, New York.

Hooper MD. (1970) Dating hedges. *Area* **2,** 63-5.

SAS Institute Inc. (1989) *SAS/STAT Users Guide*, Version 6, 4th edn, Vol. 1. Cary, NC: SAS Institute Inc. 40-3.

Smith H, Macdonald DW. (1992) The impacts of mowing and sowing on weed populations and species richness in field margin set-aside. In: Clarke J, ed. *Set-aside*. BCPC Monograph No. **50,** 117-22.

CHAPTER 10

HEDGES AS HISTORIC ARTEFACTS

D. Morgan Evans

Society of Antiquaries of London, Burlington House, Piccadilly, London W1V OHS

Introduction

Hedgerows tend to be associated with ecological and aesthetic values. It can be argued, however, that the historic aspect of hedgerows deserves more attention, not only in its own right, but also for the light that it will throw on other aspects of the subject.

It is worth beginning by stressing the obvious. There are two important practical points about hedgerows. They are created and managed by humans; and they have formed, or still constitute, boundaries to particular areas of land. Hedgerows can therefore be considered as human-made artefacts. An artefact is defined as a product of human workmanship; it is distinct from a similar object naturally produced, and is a thing not naturally present. Although the material of which a hedge is made is, almost by definition, 'natural' and is still living, the making of the hedge, and especially its maintenance, involves a high degree of human intervention and often skilled labour. Similarly, as 'boundaries', hedgerows are just one way of defining space in the countryside along with, for example, walls, banks and fences. An over-concentration on the materials of which the 'hedge boundary' is made can obscure its historic role as part of the human response to the problems of definition and control of space in the countryside. With these points in mind, this article looks first at the evidence for the dating of hedgerows and then considers how their role and importance as 'historic artefacts' can be used as an aid to their conservation.

How old are hedges?

Dating hedges from medieval times onwards using, for example, 'Hooper's Rule' is well enough described elsewhere (Hooper,1974; Rackham,1986). The techniques practised make use of the number and types of species found in the hedge, and the results can be subjected to some checking by reference to maps and documents. The methods used have their problems, but this system has been tested widely enough to give a reasonable degree of confidence that, used carefully, it can greatly assist in dating hedges back to the late Anglo-Saxon period. At the least, it can broadly indicate which hedges are 'recent' and which are probably 'old'. It should also cast light on other factors, such as which hedges have been created from woodland clearance, or which have been created by the planting of several species rather than one. Because documentary sources can to some extent be drawn upon for confirmatory evidence, we are properly within the strict definition of 'historic'.

Our problems begin when we attempt to date hedges back further than about AD 800. The numbers of species in a length of hedge do not, apparently, increase in number *ad infinitum* and we have no documentary sources dealing with boundaries before the Anglo-Saxon charters. The question, therefore, is whether there were hedges in this country before this date? In attempting to give an answer, we need first to consider the problems of identifying hedges in the archaeological record for it is to this source that we first need to turn.

Hedges and the archaeological record

The best source of evidence for a hedge would be the physical preservation of hedge remains on a waterlogged site. Most archaeological sites are not waterlogged so the chances of finding this material are obviously small. The most commonly recognized hedge species do not figure in the pollen record so this source, even where it exists, will not be helpful. These problems have been described in more detail (Robinson, 1978). The evidence of snails from prehistoric sites may provide some help, especially the occurrence of *Pomatias elegans* (Bowen *et al.*, 1978) and this deserves more study. Attempts to postulate the existence of hedges, as described below, involve the interpretation of evidence which can not be used to reach other than tentative conclusions. With these reservations in mind the sections below look at what evidence there is for hedges in Britain and Ireland in the Roman and prehistoric periods.

Roman hedges

While Rackham (1986) cites contemporary literary evidence for the existence of hedges in Italy and perhaps Flanders during the Roman period, this can not be assumed to indicate the presence of hedges at this period on our island. As Rackham indicates, however, there are two archaeological sites which suggest the presence of hedgerows in Britain in the Roman period.

Bar Hill Roman Fort

The first site is that of Bar Hill Roman Fort in Dumbartonshire (Boyd, 1984). A ditch was found underlying the Roman fort. It had been infilled prior to the construction of the fort with large turves and well preserved branches laid 'in bundles resembling armfuls of brushwood'. The pollen analysis was interpreted as showing forest clearance associated with grazing and, later, partial expansion of heather associated with declining grazing pressure. The wood was identified as *Crataegus* sp. (hawthorn). Attention was drawn to the evidence for interference in the growth of the wood some two to three years before its deposition in the ditch. The interference can be seen as consistent with hedge-laying, and therefore as possible evidence for the presence of hedges in the area. If this were the case, it could be further surmised, given the presence of the turves, that they were being used for pastoral stock control. The control of stock by hedges need not be inconsistent with declining grazing pressure elsewhere. The pollen record does not include hawthorn pollen, but this is due to the fact that hawthorn (and blackthorn) are insect-pollinated and so only rarely figure in fossil pollen deposits.

Farmoor

The second site is Farmoor in Oxfordshire. The Farmoor site has produced evidence which might suggest the presence of thorn hedges in Roman times (Lambrick and Robinson, 1979). In waterlogged deposits there were found thorny twigs of blackthorn/hawthorn type and evidence of *Rosa* (rose) bud scales and prickles. In the Roman pollen samples, tree pollen was low and other evidence suggested that the occupation of the site was at its most intensive in the late Roman period. The excavators deduced that the site had not been abandoned to scrub and that the thorn and rose combination possibly indicated the presence of thorn hedges. While the evidence is not conclusive, the likely function of such hedges would have been the enclosure of small fields to allow a more intensive and varied use of the land. It cannot be proved that arable agriculture was practised on the site, but neither can we assume its absence. The Roman evidence from this site contrasts with that of the earlier Iron Age phase where there is no indication at all of the existence of hedges.

While not strictly applicable to stockproof hedges it is worth noting that there is reasonable evidence for box hedges on Roman villa sites (Zeepvat, 1991). The evidence from Bar Hill and Farmoor, together with the documentary evidence referred to above, indicates that there may well have been hedges of blackthorn/hawthorn in Roman Britain. It might even be that the laid hedge as we know it dates from this period, since its maintenance presumably depends upon the common availability of iron cutting tools.

Studies of Romano-British iron tools shed some light on the problem. A survey of prehistoric and Roman agricultural implements in Britain (Rees, 1979) and the catalogue of the large collection of such tools in the British Museum (Manning, 1985) showed that both lighter pruning tools and heavier billhooks were available to the people of this period. The comment was made that 'differentiating between light billhooks and heavy reaping hooks is largely a matter of opinion' (Manning, 1985). There are also a large number of lighter hooks which could have been used for pruning or cutting leaves for fodder. These could equally well have been used for trimming or 'brushing' a hedge. The heavier billhook occurs less often in the archaeological record, but it is noted that both the 'modern' types of billhook were in existence apparently before the end of the Iron Age (Manning, 1985). The appearance and increase in the number of these specialist tools has been linked with the increasing importance of foliage as fodder in a time of climatic deterioration (Rees, 1979). This raises issues which deserve more detailed attention about the possible role of hedges as sources of fodder, for example, the East Anglian elm hedges (Rackham, 1986).

We can conclude our consideration of the evidence for Roman hedges with the suggestion that the continental literary evidence, the archaeological evidence and the range of tools available all point to the existence of laid hedges in Roman Britain. It should, however, be remembered that the techniques and technology for coppicing wood were in existence in the neolithic period (Coles and Coles, 1986) and that a living barrier can be formed without iron tools.

Pre-Roman hedges

There is possible evidence for one Iron Age hedge at Alcester, Warwickshire (Greig, 1992). A waterlogged pre-Roman watercourse was excavated, producing material dated to the Iron Age. The organic material was representative of various plant communities including aquatic, marsh, weeds including crop weeds, crops, and managed grassland. There were also abundant remains of scrub plants including *Crataegus* sp., *Prunus spinosa, Acer campestre, Alnus glutinosa*, and *Rosa* sp. The tentative conclusion has been drawn that the concentration of these typical hedgerow taxa at a boundary of some kind seems to show that there was a hedge there. Other than this possible site there seems to be no evidence for Iron Age hedges although, as has been pointed out above, the technology to lay such hedges existed.

In contrast to the Iron Age there are several sites dating from the Bronze Age where hedge remains might occur.

Ashville
The first site is at Ashville near Abingdon, Oxfordshire and has already been commented on elsewhere (Parrington, 1978; Fowler, 1983). At this site Middle Bronze Age cremation pits produced charcoal fragments including either hawthorn or blackthorn, with a preference towards the latter as blackthorn seeds were also found. In interpreting this evidence two possibilities arise. As a colonizing species, the blackthorn might have simply been a feature of pastoralism or transient arable; or it could have been the product of 'hedges'. The blackthorn would presumably have been already dried in order to provide the cremation fire. It is possible that hedge-trimmings were a convenient source of such kindling.

Slough House Farm
The site at Slough House Farm, near Heybridge, Essex has produced evidence which suggests that there may have been a hedged landscape in the Bronze Age which persisted through to Roman times, but by the seventh century AD had disappeared (Wiltshire, 1992). The evidence comes from the sediments of a series of waterlogged features which might have been watering holes or shallow wells. The broad interpretation of the data came from an attempt to integrate macrofossil and pollen evidence. The full publication of this site will be of special interest. The suggestion that hedges could be degraded and disappear in antiquity is always worth bearing in mind. The possibility of some relationship between early hedges disappearing by the seventh century and the limits of the Hooper dating method to late Saxon times provides a cautionary note, although the evidence is far too limited to provide other than an interesting train of thought.

Shaugh Moor
The third Bronze Age site is part of the complex landscape of Shaugh Moor on Dartmoor, and it is in this area that the relationship of botanical evidence to archaeological remains becomes most interesting (Balaam *et al.*, 1982). In the lower layers of a Middle Bronze Age ditch, a quantity of wood was recovered from waterlogged deposits. It all appeeared to be stray material accidentally incorporated in the ditch fill. Amongst the species identified were *Rosa canina* (dog rose) and

Crataegus (hawthorn) type. The hawthorn was fifteen years old. *Quercus* sp. (oak) was also present, possibly derived from some form of coppice management, as well as split oak which might have formed the bases and fragments of decayed fence posts -- perhaps of a post and rail type rather than paling. Examination of one timber led to the interpretation that it had acted as part of a gateway barrier. The excavators also suggested that the dog rose formed some sort of prehistoric 'barbed wire'. The presence of hawthorn and dog rose could also lead to the conclusion that, as in the case of the Farmoor site described above, some sort of 'hedge' may have been present on the site. The waterlogged ditch formed part of a complex boundary which, in its turn is part of the wider Dartmoor Bronze Age landscape.

The boundary of which the ditch formed a part had two phases. The first phase was interpreted as a 'post and panel' fence, occasionally flanked by a ditch and supported by a low clay mound. In the second phase the fence was replaced by a stone wall. The presence and general direction of hoofprints in the ditch of the first phase show that the boundary was stockproof, as does the need for a gate referred to above. The general sequence thus set out also applies to sites excavated on Holne Moor on Dartmoor (Balaam *et al.*, 1982) and similar sequences can be found elsewhere, for example, at Trethellan Farm, Newquay, Cornwall (Nowakowski, 1991). The sequence of an earthen bank associated with a fence line which subsequently has stones piled on it is easy enough to identify in the archaeological record, but the practical boundary and botanical implications deserve further examination.

The first point to consider is the reason for making an earthen bank. While it may provide support for the fence line, as is sometimes suggested, its provision would seem to demand an excessive use of labour, especially if the fence is of the 'post and panel' type. In the case of Trethellan Farm, the putative fence is off to one side and the excavator questions the usefulness of the first phase bank (Nowakowski, 1991). The answer is that it would be possible in all cases for this low bank to form the base for planting a hedge. The second phase of stone dumping on the bank is sometimes interpreted as a stone 'wall' which, while the stone may be sometimes laid in layers, seems often to be of a patchy nature. At Trethellan Farm part of the 'wall' shows signs of construction in one place, but in another seems to have been no more than hand-sized stones thrown up from field clearance. Rather than being a deliberately constructed wall, this phase would seem to represent the sort of conditions that can be found at the base of many hedges on earthen banks. Even where the ratio of stone to earth is high, at least one archaeologist who has studied Dartmoor in great detail has concluded that 'hedges' can form on stone banks such as those which were constructed during the Bronze Age in that area (Fleming, 1988); and it is the presence of the 'shrub' growth that clearly converts a low boundary to a potentially stockproof barrier.

In considering how hedges might have formed in this period there are two strands of argument. On the one hand there is the possibility that the low earthen banks are thrown up for hedges to be planted on them. On the other, there is the possibility that hedge formation took place along a fence line in the manner described by Rackham (Rackham, 1986). Whatever the circumstances of formation there does seem to be a serious possibility of hedges in the Bronze Age.

Neolithic hedges?

Can we look further back? It is certainly suggested that in the neolithic period the presence of stake fences found sealed by long barrows implies fixed plots and this would provide the environment for hedge formation in the manner mentioned above (Whittle, 1985). At Ceide Fields, Co. Mayo, Ireland, it is suggested that a neolithic countryside of regular rectangular fields can be identified over some ten square kilometres (Caufield, 1978, 1993). No evidence appears to exist for hedges occurring on the low stone field walls, but if the dating is correct this adds to the evidence for the presence of fixed boundaries at an early date. We also know that the ability existed to manage coppice and make hurdles (Coles and Coles, 1986), but as yet we have no evidence for neolithic hedges.

On the evidence available it is suggested that in the Roman and Bronze Age periods there is a reasonable possibility that hedges existed in Britain. While the site at Slough House Farm provides the suggestion that hedges could have been lost in antiquity, the question that now needs to be addressed is whether any early pre-Saxon hedges could have survived through to the present day?

The possible answer to this lies in the technique of analysing field patterns known as topographic analysis. This technique looks at field patterns and tries to take them back to their original forms. Boundaries are also investigated to see if their physical characteristics allow them to be differentiated. Two good examples of this approach, in differing locations and soils, can be found for East Anglian Boulder Clays (Williamson, 1987) and for Derbyshire (Wildgoose, 1991). The important conclusion from these, and other studies, is that extensive areas with pre-Roman boundaries still intact do survive in England. If the continuity of use implicit in this process has ensured the survival of patterns of boundaries, this suggests the possibility of significant amounts of hedgerows with prehistoric origins also surviving. If this conclusion is correct there is clearly the need for more research work to determine why there are the apparent limits to the Hooper Rule and whether botanical techniques can help identify and date Roman and prehistoric hedges.

Conclusion

The conclusion drawn from this section is that there is a reasonable possibility that extensive lengths of hedges of Bronze Age and Roman origin exist today in the countryside. When we turn to consider the relative historic importance of hedges this is an obvious factor. What is also very important to remember is that the undoubted existence of hedges from about AD 800 remains unquestioned.

The future of hedges

In looking at the future of hedges two approaches must be considered. The first involves the continuing maintenance of hedges which is necessary if they are to be kept in good condition. Maintenance requires physical and financial resources and, if the hedge is to be maintained in an optimum condition, specialist skills. This approach clearly needs the provision of training and grants. While there may be reservations about the level of grants, the Ministry of Agriculture hedgerow element of their Farm and Conservation Grant Scheme and the Countryside Commission's Hedgerow Incentive Scheme provide a recognized framework for resources for management (Ministry of Agriculture, Fisheries and Food, 1992). The second approach to be considered is the need to control the unecessary destruction of hedges. This destruction currently seems to be more from building development, golf course construction, mineral extraction, and road construction than from agricultural fashions. These processes of destruction all lie within planning control mechanisms. The question, then, is whether hedges as 'historic artefacts' can be regarded as 'material considerations' in the planning process.

Hedges in the planning process

The nature of hedges as 'human artefacts', as defined above, suggests that one approach to seeing what role hedges might play in the planning process may be found in the areas of definition of the 'built environment', rather than those devoted to 'natural history', although as will be described, there is perhaps not as great a difference between these two as might be expected.

The starting point for regarding hedges as material considerations in the planning process is the Department of the Environment Planning Policy Guidance Notes (PPG's). PPG 12 'Development Plans and Regional Planning Guidance' (Department of the Environment, 1992*b*) sets the overall tone on Development Plans and Regional Planning Guidance. In the Introduction, paragraph 6.3, Local Planning Authorities are told to take account of the environment in the widest sense in preparing plans and they should be familiar with the issues of, amongst other matters, 'the built heritage'. Amongst the policies for land use are 'giving high priority to conserving the built and archaeological heritage' (paragraph 6.5). This is coupled with the strong value that people place on their environment including 'landscape conservation and the built heritage'.

The next Planning Policy Guidance Note to be considered is PPG 7, 'The Countryside and the Rural Economy' (Department of the Environment, 1992*a*). This states specifically that 'new development in rural areas should be sensitively related to ... the historic, wildlife and landscape resources of the area' (paragraph 1.10). This is reinforced by the need, when preparing Development Plans, 'to protect landscape, wildlife and historic features' (paragraph 2.4). It is also worth noting that in Environmentally Sensitive Areas 'the features which contribute to the designation of the area ... may sometimes also be important features in local countryside planning'.

If hedges are accepted as historic artefacts, it would seem that they could have a clear role as factors in the planning process. Do they also have a role as 'archaeology'. The PPG which deals with 'Archaeology and Planning' is number 16 (Department of the Environment, 1990*b*).

In PPG 16 much hinges on the use of the words 'archaeological remains' in paragraph 6. These words are the main description of the subject matter to which the PPG refers. These 'archaeological remains' are not clearly defined, but are described as 'finite and non-renewable ... in many cases highly fragile and vulnerable to damage and destruction. Apppropriate management is therefore essential to ensure that they survive in good condition'. In paragraph 4, these 'archaeological remains' are seen in terms of 'today's archaeological landscape' which is 'the product of human activity over thousands of years'. The remains of the activity date from 400,000 years ago to the early twentieth century and includes 'farms and fields'. It is doubtful if those who drafted PPG 16 saw hedges as being included amongst 'archaeological remains' but it could be, perhaps should be the position that, in appropriate circumstances, hedges are included. English Heritage certainly recognize that hedges are part of the historic component of the landscape. In their statement on historic landscapes they state that as historic components hedges form part of 'the very grain of the land' (English Heritage, 1991), and future proposals on an Historic Landscapes Register, requested by Government (Department of the Environment, 1990*a*) are awaited with interest even if such a register were to be non-statutory.

If hedges are to play a role in the planning process and if there are to be resources to manage them it is necessary that some method of distinguishing the relative values that can be put on different hedges be agreed.

Valuing historic hedges

It is axiomatic that not all hedges are of equal historical interest or importance, and it should be possible to define a basis for judging this. This is needed, not for academic purposes alone, but also, given the role that hedges can play in wider countryside management and the human and financial resource implications of this, as an aid to public policy making. Any system which seeks to define the historic importance of hedges for reasons of public policy needs to have a framework which is clearly defined and open to critical examination.

Judging a hedge's historic importance

'Human artefacts', in the form of 'ancient monuments', can have their 'national importance' defined within a framework set out by non-statutory criteria, for example, those issued most recently in the Department of the Environment Planning Policy Guidance Note 16, Planning and Archaeology (PPG 16), Appendix 4. These criteria were originally issued to apply simply to 'ancient monuments' of sufficient importance to be scheduled under the Ancient Monuments and Archaeological Areas Act 1979. Their appearance in a Planning Policy Guidance Note clearly extends their role, because references are made to 'archaeological remains' which are clearly different from, in the sense of being more widely drawn than, 'ancient monuments' which are precisely defined in legal terms (Ancient Monuments Act 1979, Section 61 (1),(7) and (12)). These 'archaeological remains' are also capable of being of national importance and the conclusion must be reached that the Department of the Environment and, presumably, the Department of National Heritage, see them as having greater importance in future. The point needs to be emphasized that it is Government which has officially extended the use of these criteria to well beyond their original subject.

The criteria set out in PPG 16, Appendix 4, are listed, but it is important to look first at the introductory paragraph. The use of the criteria is described as a decision-making framework 'for assessing ... national importance', and this element is brought out in the first paragraph. The list of 'criteria' is not *definitive*, but they are rather seen as *indicators* 'which contribute to a wider judgement based on individual circumstances' and they are also not given in any order of ranking.

We can now look at the list of criteria to see if they really can be applied to hedges in the same way that they are to archaeological remains or monuments. The one that first comes to the fore is:

(v) Survival/condition, stating that the remains should 'be assessed in relation to its present condition and surviving features'. This can obviously be applied to hedges where the state of a hedge is an important part of recognizing and defining its general value.

Next there are some relatively straightforward criteria, all of which can be applied to hedges:

(i) Period, 'all types ... that characterize a category or period should be considered for preservation' which can be applied, as we will see, to Enclosure period hedges as much as to Saxon ones.

(ii) Rarity, 'some ... categories ... are so scarce' but 'a selection must be made which portrays the typical and commonplace' to which the comment made on 'period' can be applied.

(iii) Documentation, remains whose 'significance ... may be enhanced ... by the existence of written records'. The presence of documentation in the form of Saxon Charters or even Enclosure Acts is applicable.

(iv) Group value, 'the value of a single monument may be greatly enhanced by its association with related contemporary monuments or with monuments of different periods'. The approach to field pattern analysis described above obviously applies here. The approach can be extended to other archaeological and historic remains, including buildings.

(vi) Fragility/vulnerability, can be considered as self-evident for hedges as a class of artefact 'whose value can ... be severely reduced by neglect or careless treatment...'. This would not apply in a particular case - for example, where the hedge is in very poor condition. The remaining criteria are **(vii) Diversity** and **(viii) Potential,** which are probably less helpful for the selection of hedges.

Before looking at how these criteria might work in practice, it is worth considering the links of the PPG 16 criteria with other criteria. The most obvious parallel is with the 'Ratcliffe' criteria as set out in in the *Nature conservation review* (Ratcliffe, 1977). The critera are listed as **Diversity, Naturalness, Rarity, Fragility, Typicalness, Recorded History, Position in an ecological/geographical unit, Potential value,** and **Intrinsic appeal**. The links with the PPG 16 criteria are self- evident and it is suggested that, in embryonic form at least, there exists a common approach to assessing the importance of natural and human-created features in the countryside - the more so since in this country the human influence on the 'natural' environment is so significant.

What are the practical problems which might arise from using the PPG 16 criteria to judge the importance of a hedge? The first must be one of the identification of species which is to an extent linked to seasonal appearance and also to the ecological range, present and past, of individual species. Added to this is the sheer amount of effort that is required in walking the lengths of hedges and recording the ground-based information. The second problem is gaining access to documentary material. In the case of archaeological sites a large database already exists based largely in County Councils. While this is by no means complete it does give a good start. No such database at the moment exists for hedges. This is perhaps something that should be addressed. The ability to store, access and control information on machine-based systems means that the amount of eventual data is not quite the obstacle that it might once have been.

Assuming that we have all the relevant information about a hedge what implications are there in using the PPG 16 criteria? We have already commented on the primary role of **Condition**. It should be used with care, however. A hedge in a poor condition, low in species and very gappy could nonetheless be of great importance because of its documented history, say as a Saxon Charter boundary. **Rarity** is a relatively easy concept to grasp, but the criteria do say, as mentioned above, that 'a selection must be made which portrays the commonplace as well as the rare'. If this approach is taken with hedges it would mean that we would have to tackle the vexed question of Enclosure hedges. The view has been taken, and received a wide airing, that 'there is little to be said for trying to preserve the typical Enclosure Act hedge' (Rackham, 1986) This is said by comparison with ancient hedges. However, if we find an area of 'typical' Enclosure hedges which are in good and well-maintained **Condition**, which we know have an extensive **Documentation** and where not only the complete extent of the system survives but also the contemporary buildings give a high **Group value,** then it would be suggested that here we have something of high historical value. Age as a measure of value is not always the most helpful approach to judging the historical importance of something. For example, 'historic buildings' which are less than thirty-years-old can be protected by being giving a 'listing'. In multi-period landscapes Enclosure hedges can also clearly be of **Group value** where they fossilize the boundaries of an open-field system. Some of these boundaries may be of an earlier date.

Conclusion

In conclusion it is suggested that hedges can be treated as 'artefacts' and that this brings them legitimately into the field of study of the archaeologist. While archaeological excavation techniques might provide information on the dating of early hedges it would appear that they need to be combined with techniques of topographic analysis to give results of real use. This area of study would benefit from more research. Present indications are that hedges, as a distinct feature, can certainly be dated to the late Saxon period, probably to the Roman period and possibly to the Bronze Age. Hedges can properly be seen as part of the 'built environment' and could be incorporated into the existing planning system as material considerations. Existing guidelines such as PPG 16 would appear to offer an adequate set of criteria for making professional judgements about the relative archaeological and historical importance of hedges. The parallels between archaeological and natural history criteria offer the opportunity of developing a joint approach to valuing hedges, recognizing both strands.

References

Balaam ND, Smith K, Wainwright GJ. (1982) The Shaugh Moor project: fourth report-- environment, context and conclusion. *Proceedings of the Prehistoric Society* **48,** 203-78.

Bowen HC, Evans J, Race E. (1978) An investigation of the Wessex Linear Ditch System. In: Bowen HC, Fowler PJ, eds. *Earlyland allotment in the British Isles.* British Archaeological Reports **48,** 149-54.

Boyd WE. (1984) Prehistoric hedges: Roman Iron Age hedges from Bar Hill. *Scottish Archaeological Review* **3,** Part 1, 32-4.

Caufield S. (1978) Neolithic fields: the Irish evidence. In: Bowen HC, Fowler PJ, eds. *Early land allotment in the British Isles.* British Archaeological Reports **48,** 137-44.

Caufield S. (1993) *Ceide Fields, Ballycastle, Co.Mayo.*

Coles B, Coles J. (1986) *Sweet track to Glastonbury.*London: Thomas and Hudson.

Department of the Environment (1990*a*) *This common inheritance.* London: HMSO.

Department of the Environment (1990*b*) Archaeology and planning. *Planning Policy Guidance* **16**. London: HMSO.

Department of the Environment (1992*a*) The countryside and the rural economy. *Planning Policy Guidance* **7**. London: HMSO.

Department of the Environment (1992*b*) Development plans and regional planning guidance. *Planning Policy Guidance* **12**. London: HMSO.

English Heritage (1991) The historic landscape: an English Heritage Policy Statement. *Conservation Bulletin* **14,** 4-5.

Fleming A. (1988) *The Dartmoor Reaves.* London: Batsford Ltd.

Fowler PJ. (1983) *The farming of prehistoric Britain.* Cambridge: Cambridge University Press.

Greig JRA. (1992) *An Iron Age hedgerow flora from Alcester, Warwickshire.* English Heritage, Ancient Monuments Laboratory Report 11/92.

Hooper MD. (1974) Dating hedges. *Area* **4,** 63-5.

Lambrick G, Robinson M. (1979) *Iron Age and Roman riverside settlements at Farmoor, Oxfordshire.* Council for British Archaeology Research Report **32**.

Manning WH. (1985) *Catalogue of the Romano-British iron tools, fittings and weapons in the British Museum.* British Museum Publications Ltd.

Ministry of Agriculture Fisheries and Food (1992) *Conservation and diversification grants for farmers.* London: HMSO.

Nowakowski JA. (1991) Trethellan Farm, Newquay: the excavation of a lowland Bronze Age settlement and Iron Age cemetery. *Cornish Archaeology* **30,** 5-242.

Parrington M.(1978) *The excavation of an Iron Age settlement, Bronze Age ring-ditches and Roman features at Ashville Trading Estate, Abingdon (Oxfordshire) 1974-76.* Council for British Archaeology Research Report **28**.

Rackham O. (1986) *The history of the countryside. London:* J.M.Dent & Sons Ltd.

Ratcliffe DA, ed. (1977) *A nature conservation review*. Cambridge: Cambridge University Press.

Rees SE. (1979) *Agricultural implements in prehistoric and Roman Britain*. British Archaeological Reports **69**.

Robinson M. (1978) The problem of hedges enclosing Roman and earlier fields. In: Bowen HC, Fowler PJ, eds. *Early land allotment in the British Isles*. British Archaeological Reports **48**, 155-8.

Whittle A. (1985) *Neolithic Europe: a survey*.Cambridge: Cambridge University Press.

Wildgoose M. (1991) The drystone walls of Roystone Grange. *Archaeological Journal* **148**, 205-40.

Williamson T. (1987) Early co-axial field systems on the East Anglian boulder clays. *Proceedings of the Prehistoric Society* **53**, 419-32.

Wiltshire PEJ. (1992) *Palynological analysis of sediments from a series of waterlogged features at Slough House Farm, near Heybridge, Essex*. English Heritage, Ancient Monuments Laboratory Report **25/92**.

Zeepvat RJ. (1991) Roman gardens in Britain. In: Brown AE, ed. *Garden archaeology*. Council for British Archaeology Research Report **78**, 53-9.

CHAPTER 11

HEDGEROWS: LINKING ECOLOGICAL RESEARCH AND COUNTRYSIDE POLICY

C. J. Barr[1] and T. W. Parr[2]
[1] *Institute of Terrestrial Ecology, Merlewood Research Station, Grange over Sands, Cumbria LA11 6JU, UK*
[2] *Department of the Environment, Directorate of Rural Affairs, Tollgate House, Bristol, UK.*
Current address: *Institute of Terrestrial Ecology, Monks Wood, Abbots Ripton, Huntingdon, Cambs PE17 2LS, UK*

Introduction

On 25 July 1991 Mr Tony Baldry, Junior Environment Minister, announced to the UK Parliament proposals to protect key hedgerows in England and Wales. The statement he made came at the culmination of persistent pressure from the public and non-government organizations over loss of hedgerows in the countryside and was influenced by results from research commissioned by the Department of the Environment.

In this paper we outline some of the key stages in the process of hedgerow policy development together with the key research inputs which went into this process. We present this as a successful example of the way in which the political process and ecological research and monitoring at the science/policy interface can go hand in hand in the development of countryside policy.

Development of hedgerow protection policy in the 1980s

Some of the key events in the development of hedgerow protection policy over the past 10 years are shown in Table 11.1. Although policy development, research and public opinion were inextricably linked during this period, it is probably correct to interpret public opinion and concern over perceived losses of hedgerows, fuelled by pressure from non-government organizations such as the Royal Society for the Protection of Birds (1988) and the Council for the Protection of Rural England, as the driving forces behind the process of policy development. As an example, during the ten years from 1982 to 1992 there was a total of 90 parliamentary questions asked on the subject of hedgerows (Table 11.2). Questions on the lengths of hedgerows and losses were common in the early 1980s, but despite local studies showing hedgerow losses since the war and a general public perception of a changing countryside, at that time there were no national data on hedgerow losses.

Table 11.1 Some key events in the recent history of hedgerow protection policy

Date	Policy/background	Public input	Research
1985	MAFF grants Private Member's Bill on hedgerows	Pressure from parliamentary questions, non- government organizations and press	
1986			Monitoring Landscape Change (MLC) - land cover survey results Public attitudes survey
1989	Private Member's Bill on hedgerows (Hardy)		Public attitudes survey
1990	The Batho Report - review of Tree Preservation Orders		Countryside Survey 1990 - work begins
	'This common inheritance'- Environment White Paper promises hedgerow protection measures		ITE report on 'Hedge Management' for DOE
	Consultation paper	Wide consultation on TPOs and hedgerows	
1991	Planning and Compensation Bill (proposed amendment to include hedgerow protection)	Pressure and lobbying from NGOs for hedgerow protection	
	Ministerial statement in July announces: Hedgerow Incentive Scheme		
	Hedgerow Notification Scheme		
	Ministerial statement in October on hedgerow losses 1984-1990		ITE report on 'Changes in Hedgerows 1984 to 1990' for DOE.
1992	July - Countryside Commission launch Hedgerow Incentive Scheme		ITE report on 'Hedgerow Diversity' for DOE
	October - Private Member's Bill on hedgerow protection (Ainsworth)		

Table 11.2 Parliamentary Questions (PQs) on hedgerows 1982-1992
(some questions covered more than one subject so the categories used are not mutually exclusive).

	Number of questions asking for:					
	Action	Information	Statistics on lengths/loss	Statistics on grants	Other	Total PQs
Session						
1992/93	3	-	-	-	-	3
1991/92	5	1	-	-	1	7
1990/91	2	3	2	-	-	7
1989/90	2	1	6	-	1	10
1988/89	4	3	5	3	3	18
1987/88	8	3	4	3	3	19
1986/87	-	1	1	2	2	4
1985/86	2	1	1	-	1	5
1984/85	-	-	1	1	2	4
1982/83	2	4	1	4	2	13
TOTAL	28	17	21	13	15	90

Perhaps the first national overview of hedgerow losses since the 1940s was provided by the *Monitoring landscape change* survey (Hunting Surveys and Consultants Ltd., 1986). This used aerial photographic interpretation of a stratified sample of land in England and showed a loss of 22% of hedgerows between 1947 and 1985 (Table 11.3). Hopes that these results would be regarded as definitive and lead to immediate action on hedgerow protection were short-lived as MAFF pointed to contradictory findings from their own survey work - as shown in this reply to a parliamentary question on 18 April 1990 made by a Minister at the time, Mr Heathcoat-Amory:

'A survey of landscape change in England and Wales carried out for my Department and the Countryside Commission in 1986 by Hunting Surveys and Consultants Limited showed annual losses in England and Wales of some 2,600 miles of hedgerow between 1947 and 1969; 2,900 miles between 1969 and 1980; and 4,000 miles between 1980 and 1985. A postal survey of farmers on environmental topics carried out by the Ministry of Agriculture, Fisheries and Food in 1985, however, showed annual losses of 500 miles between 1980 and 1985. The two surveys used different methodologies and neither can be regarded as definitive...'

Table 11.3   Length of hedgerows in England and Wales from 1947 to 1985
Data derived from aerial photographs in *Monitoring landscape change* survey (1986).

Region	Length of hedgerows ('000 km)			
	1947	1969	1980	1985
South East	117.9	98.7	92.8	88.3
South West	119.5	111.7	105.4	103.3
East Anglia	61.1	53.2	46.6	43.1
East Midlands	86.6	73.0	65.9	60.9
West Midlands	79.0	67.3	63.0	52.8
North West	27.9	23.9	23.4	22.1
Yorks/Humberside	131.2	112.0	101.3	96.7
Northern	39.3	37.8	35.8	33.7
Wales	133.6	125.3	118.5	114.1
TOTAL	796.1	702.9	652.7	621.0

Despite this lack of agreement over the magnitude of hedgerow losses in the early eighties public concern continued to mount. The Department of the Environment's 'Public attitudes to the environment survey' (Department of the Environment, 1989) showed that the proportion of people 'worried' to some degree by the 'loss of hedgerows' was 52%, while in 1986, 72% of people were concerned about the 'loss of trees and hedgerows'.

In response to this mounting concern, the Government included the issue of hedgerow protection in its review of tree preservation policies and legislation (Batho, 1990). This review pointed out that, in contrast to trees, hedgerows needed active and continuing management if they are to stay in good condition and that therefore controls should not be aimed solely at preventing loss. The report recommended that local authorities should have powers to make Hedgerow Management Orders to ensure the maintenance of important hedges.

A Government commitment to act on this recommendation came soon after the review in its Environment White Paper, '*This common inheritance*' (Department of the Environment, 1990). It promised:

'... to enable local authorities to protect hedgerows of key importance by making preservation orders, with appropriate payments to farmers to look after them properly; ...'

The White Paper was quickly followed by a consultation paper in which the Government outlined its proposals for Hedgerow Management Orders to secure the management of important, but threatened, hedgerows. Management grants were proposed to cover the costs of hedge cutting and laying over a twenty-year period.

Following the publication of this consultation paper in December 1990, a number of events occurred which influenced the final form of the proposals made by the Minister in his statement of July 1991:

- First, although the overall response to the consultation paper was favourable, there were some valid criticisms, particularly over the restricted range of management options suggested in the paper and also because management grants were targeted only at those hedgerows under threat rather than key hedgerows in general.
- Second, a research report on hedge management produced by the Institute of Terrestrial Ecology (Hooper, 1992) summarized the wildlife and environmental value of hedges, particularly where these were allowed to grow uncut within a regular long-term cycle of rejuvenation by coppicing or laying (Table 11.4). This work supported the idea that since there was little in the way of environmental benefit from hedge cutting, management grants should not usually include the costs of annual cutting but should concentrate on techniques aimed at rejuvenating hedgerows including laying, coppicing and the infilling of gaps.

Table 11.4   Environmental benefits of main hedge types
(measurements give typical hedge height x width)

	Hedge type			
	Unmanaged	2-sided 4m x 2m	Tall-flailed 2m x 1.5m	Short-flailed 1m x 0.75m
Relative values for:				
Small mammals	++	+	+	?
Birds	+++	++	+	?
Invertebrates	++++	+++	++	+
Herbaceous plants	+	+	+	+
Landscape	++++	+++	++	+
Sport	++++	+++	++	+
Amenity	+++	++	+/-	?

(Adapted from Hooper,1992)

- Third, although the Government was already committed to hedgerow protection, an attempt to force the pace was made through a proposed amendment to the Planning and Compensation Bill, which would have allowed local authorities to protect important hedgerows. This amendment was only narrowly avoided, but as a result the Government became committed to making a full statement containing its proposals for hedgerow protection before the end of the 1990/91 session in July.

- Lastly, there were still no definitive data on hedgerow change and critics of the Government's hedgerow policy were able to suggest that hedgerow destruction was a thing of the past.   However, early indications from the results of *Countryside Survey 1990* (Barr *et al.*, 1993), a comprehensive survey of the British countryside, revealed that not only were hedgerows still being removed, but that there was also an escalating problem of lack of hedgerow management. These results were particularly significant in maintaining the momentum for hedgerow protection measures and are therefore discussed in more detail in the following section.

Change in hedgerows in Great Britain 1984-1990 - results from *Countryside Survey 1990*

Introduction to the Institute of Terrestrial Ecology surveys

Countryside Survey 1990 was the third in a series of surveys associated with land use and vegetation in Great Britain.   The work was led by the Institute of Terrestrial Ecology and was jointly funded by the Department of the Environment and the Natural Environment Research Council, with additional support coming from the former Nature Conservancy Council.

The first of the Institute of Terrestrial Ecology surveys took place in 1978 (Bunce, 1979) and the primary purpose was to collect information on vegetation and soils; the survey used a sampling approach based on the Institute of Terrestrial Ecology Land Classification (Bunce *et al.*, 1983).   A secondary activity was the collection of land cover and landscape feature information from each of the 256 one-kilometre sample squares.   This included the mapping of 'hedges' as a field boundary type.

In 1984, the Institute of Terrestrial Ecology completed a repeat survey of the 256 one-kilometre squares and also surveyed a further 128 squares, increasing the sample number to 384.   The survey was designed to answer questions on land use issues and so concentrated on land cover and landscape feature mapping, rather than data collection at the detailed quadrat level of the previous survey.   Records on hedgerows were made using combinations of attributes to describe each boundary length.   The field methodology is given in Barr *et al.* (1985).

Information collected on hedgerows in the 1977/78 survey was not sufficiently detailed to make conclusions about subsequent changes in the condition or management of hedges.   However, by comparison with the 1984 data, it was possible to identify those boundaries which had been classified as hedges and which had been established, or removed, between the two survey dates. Using the results from the sample squares, estimates were derived for GB and for major regions within it.   These are described in Barr *et al.* (1986) and are summarized in Table 11.5.

Table 11.5 Hedgerow gains and losses between 1978 and 1984 (km)

	Hedgerow gain	Hedgerow loss
England	3,200	22,300
Scotland	<100	3,300
Wales	400	2,600
Great Britain	3,600	28,200

In 1990 the sample number was again increased, resulting in 508 rural squares being visited, with an additional 25 urban squares being surveyed as part of a separate study (Barr, 1990). The field survey was part of the larger project, *Countryside Survey 1990*, which also contributed to work being undertaken at the Institute of Terrestrial Ecology, Monks Wood (co-funded by the Department of Trade and Industry and the British National Space Centre) to produce a land cover map of GB from satellite imagery. Surveys of soils and freshwater biota in the Institute of Terrestrial Ecology squares were also included in the work programme. As part of the field survey, hedgerows were mapped in an identical way to methods used in the 1984 survey.

The handling of data recorded during 1990, and subsequent analysis, was scheduled to be completed during 1992/93. However, given the political interest in countryside matters, and particularly in hedgerows, analysis of the hedgerow data was brought forward and considered in isolation from all other surveyed information.

Field survey methods

A full description of the field survey methods is given in a Field Handbook (available by arrangement through the Institute of Terrestrial Ecology). The methods follow closely those used in the 1984 Institute of Terrestrial Ecology survey. The following paragraphs summarize only those methods which are relevant to this paper.

In summary, the Institute of Terrestrial Ecology surveyed the 384 one-kilometre squares which had first been visited in 1984, and mapped boundary features. Each length of boundary was mapped using OS 1:10,000 scale maps enlarged to about 1:7,000, and described using a combination of codes which referred to species, physical characteristics such as height and management features including 'gappiness' and degree of trimming (Barr *et al.*, 1991). This boundary information was mapped on a separate page from other field data, as shown in Appendix 11.1.

Boundaries were mapped and coded as 'single lines' on the map, even though there may have been several different elements associated with each (e.g. a hedge and a fence on top of a stone bank). For adjacent lines to be mapped individually, then a clear gap between all elements of the two boundaries, or an intervening feature, such as a stream, had to be identified.

The length of each boundary, or boundary segment, was determined by the constancy of a combination of codes, along the length; where any one description differed, then a new length was demarcated and a new combination of codes was used. The minimum length of boundary to be described was 20 metres and the ends of each length were marked using 'tic' marks at right angles to the mapped feature. The same coded descriptions were used in both 1984 and 1990 except that additional codes for 'regrowth from stumps' and, on another page of the recording booklet, 'line of shrub', were introduced in 1990.

To assist in field mapping, limited aerial photographic interpretation was carried out for each square. Using photographs of various dates, but all taken since the 1984 survey, boundaries that were no longer present, and those that were new to the map, were marked on a 'master map' which was used as a base for field recording.

Boundaries of land associated with buildings (curtilage) were not mapped in detail, nor were boundaries within woodland mapped.

Definitions associated with boundaries

A **boundary** in this context is defined as a physical barrier, having a height and width, usually intended to prevent farm stock from moving from one area to another. A **hedge** is a boundary, or part of a boundary, which comprises a row of bushes or low trees growing closely together, and which have been managed through cutting to maintain a more or less dense, linear barrier. **Hedgerow** is used interchangeably with hedge, although more strictly it should be used as a broader term, encompassing other features associated with hedges, such as trees and gates etc. (Hooper, 1968). Only hedgerows occurring in rural situations were considered in this work.

It can be difficult to distinguish between unmanaged hedgerows and lines of trees. When hedge management is abandoned and the natural shape of the tree is regained, then the feature can no longer be described as a hedge and is likely to be coded as a line of trees (on the trees/woodland/forestry page of the recording booklet).

The codes used in describing hedgerows and their definitions are given in Table 11.6.

Data entry and analysis

All mapped linework was digitized using ARCINFO GIS software. To ensure spatial integrity, the 1990 data were digitized and labelled first, and then each coverage was copied, edited, and re-labelled with 1984 information. This minimized technical differences such as boundary mis-matches and overlaps during overlaying (Howard and Barr, 1991).

Table 11.6   Definitions of codes used in *Countryside Survey 1990* (based on definitions given in the field handbook but using additional information resulting from a consensus agreement of the use, or limitations to use, of the code during the 1990 field survey)

Hedge>50% hawthorn	where hawthorn constitutes more than half the length of hedge under consideration
Hedge>50% other	where a species other than hawthorn constitutes more than half of the hedge, the species being recorded
Mixed hedge	used for any length of hedge where no single species dominates
Hedge trimmed	signs of management within the previous 12-24 months and a neat, cropped appearance
Hedge uncut	has had recent management but has been 'let go' over more than two seasons
Hedge derelict	still obviously a hedge but all attempts at management have been abandoned
Line of relict hedge	usually a line of trees or shrubs showing where a hedge has once been (can be used in addition to codes on the forestry page)
Laying	used if it appears likely that the hedge has been layed in the last five years
Flailing	used if flailed in the last year; recognizable by smashed and shattered ends to cut branches
Burnt	used to indicate that the hedge has been burned, accidentally or otherwise.
Regrowth from cut stumps	applies to hedges that have been cut to ground level but have grown again, often at intervals along the old boundary
Boundary>2m high or Boundary<1m high	the height codes apply to the height of the hedge at the time of survey. If different heights apply on either side of the boundary, then the code should refer to the side on which stock are kept; otherwise, the lowest height should be coded
Boundary stockproof	applies to the stock that would normally use the surrounding fields; if type of stock not clear, then assumed to be sheep
Boundary not stock-proof	
Boundary with filled gaps <10% or filled gaps >10%	used to show that the boundary has had gaps which have been filled in an attempt to make it stockproof (e.g. by short lengths of wooden fence). The percentage of gaps is of the boundary unit being coded.
Signs of replacement	used where there is evidence that one boundary type has been replaced by another (e.g. fence replacing hedge)
Signs of removal	used where there is clear evidence of boundary removal, e.g. grubbed-out hedge
No longer on map	used to indicate that a line on the OS base map no longer marks the position of a boundary on the ground

All data codes were punched twice, cross-checked, edited and a single version entered into an Oracle Database Management System which could be integrated with the digitized data.

The 1990 coverages were compared with those from 1984 using modified ARC overlaying procedures. Analysis was carried out on all boundaries to which a hedge code had been ascribed (except for relict hedges which were treated as a separate boundary type) even though other boundary features, such as walls and fences, may have contributed to the boundary (see Field survey methods on page 125).

By calculating mean boundary change statistics for each of the Institute of Terrestrial Ecology Land Classes (and knowing the frequency of the Land Classes in GB), national estimates were derived. The sample size allows such estimates to be made for large regions only - estimates for, say, smaller counties are accompanied by large statistical error terms.

Field survey results

The results of the comparisons of boundaries which contained a hedge component in 1984 and/or in 1990 are given in Table 11.7. This table shows that the net decrease between 1984 and 1990 amounts to nearly one-quarter of the length of 1984 boundaries which contained hedges. However, the net change is a balance of gains and losses, and details of these are presented in Table 11.8. (Estimates for GB, and totals, are derived separately from estimates for each country.)

Table 11.7   Estimates of net change in hedgerow lengths in England, Scotland, Wales and Great Britain between 1984 and 1990 (lengths and standard errors ($\pm$) in '000 km)

	England	Scotland	Wales	GB
Total hedge length in 1984	440.7	48.5	74.7	563.8
	($\pm$26.1)	($\pm$8.6)	($\pm$8.4)	($\pm$32.9)
Total hedge length in 1990	344.3	35.2	53.8	433.3
	($\pm$22.8)	($\pm$7.0)	($\pm$6.9)	($\pm$28.9)
Net decrease between 1984	-96.4	-13.3	-20.8	-130.5
and 1990	($\pm$12.3)	($\pm$2.7)	($\pm$3.9)	($\pm$14.9)

Table 11.8 gives estimates of the lengths of hedges that have been planted as well as those that have been removed. In addition, some boundaries have changed in their nature and appearance leading to increases and decreases in boundaries that can be defined as hedgerows. For example, lines of immature trees that have been thinned out and then laid as hedges, will lead to an increase in the estimate of hedgerow length. Conversely, where a former hedge has been unmanaged over a number of years, it will grow into a line of trees (a relict hedge). Other examples of change in boundary type include where a hedge has become 'gappy' and has been

recorded as a line of shrubs, and not a hedge, and where vegetation growing on the top of a bank has been cut in such a way that a hedge is formed.

Boundaries that were recorded as hedges for the first time in 1990 (other than those resulting from change in boundary type) totalled 11,600 kilometres. Complete removal of hedgerows between the two dates amounted to 63,900 kilometres, or 11.3% of the total 1984 hedgerow length. This paper does not consider in any detail how or why these hedgerows have been removed, or whether the loss is balanced by new planting.

Close inspection of the results shows that most change is associated with management of hedgerows. About 118,400 kilometres, or 21% of the 1984 hedgerows in GB were coded in 1990 as a different type of boundary (e.g. lines of trees or shrubs, or as relict hedgerows). Conversely, only some 41,200 kilometres of 'new' hedges in 1990 came from the re-definition of boundary types. This suggests that hedgerows were subject to less active management in 1990 than in 1984.

Table 11.8   **Estimates of hedgerow gains and losses in England, Scotland, Wales and Great Britain between 1984 and 1990 (lengths and standard errors ($\pm$) in '000 km)**

	England	Scotland	Wales	GB
1990 hedges gained				
New hedges	9.7	1.0	1.2	11.6
	($\pm$1.4)	($\pm$0.1)	($\pm$0.1)	($\pm$1.5)
Change in boundary type	20.0	2.4	4.0	26.4
	($\pm$2.4)	($\pm$0.1)	($\pm$1.0)	($\pm$2.9)
Management of relict hedges	11.5	1.0	2.4	14.8
	($\pm$2.9)	($\pm$0.1)	($\pm$1.0)	($\pm$3.3)
TOTAL GAIN	41.1	4.0	7.6	52.8
	($\pm$4.6)	($\pm$1.0)	($\pm$1.5)	($\pm$5.4)
1984 hedges lost				
Hedges removed	52.7	3.6	7.6	63.9
	($\pm$4.7)	($\pm$1.0)	($\pm$1.2)	($\pm$5.4)
Change in boundary type	51.2	9.8	14.6	75.5
	($\pm$6.9)	($\pm$1.7)	($\pm$2.7)	($\pm$9.1)
Unmanaged (lost to relict)	33.0	3.8	6.1	42.9
	($\pm$4.2)	($\pm$1.3)	($\pm$1.4)	($\pm$5.2)
TOTAL LOSS	136.9	17.2	28.2	182.3
	($\pm$10.9)	($\pm$3.1)	($\pm$3.8)	($\pm$13.7)

Examination of Tables 11.7 and 11.8 show that there are differences in hedgerow distribution between England, Scotland and Wales. For example, hedgerows in Scotland and Wales appear to have undergone more change, proportionally, than those in England (Table 11.7), but the proportion of hedges that have been removed is less (Table 11.8).

Table 11.9 gives the results of an analysis of data on 'lines of relict hedges' (defined as 'a line of shrubs or trees showing where a hedge has once been') which have been estimated independently from the hedgerow data, for 1984 and 1990. As stated earlier in this paper, many former hedgerows were re-defined as lines of trees and shrubs in the 1990 survey. The figures in Table 11.9 support the contention that a relaxation of hedgerow management has led to an overall decrease in hedgerow length and a corresponding increase in lines of trees and shrubs.

Table 11.9   Estimates of lengths of 'lines of relict hedgerow' in England, Scotland, Wales and Great Britain for 1984 and 1990 (lengths and standard errors)

	England	Scotland	Wales	GB
1984	41.4	5.2	11.3	57.8
	($\pm$4.8)	($\pm$1.4)	($\pm$2.7)	($\pm$6.6)
1990	65.1	8.1	15.3	88.5
	($\pm$6.7)	($\pm$2.2)	($\pm$3.6)	($\pm$9.4)

An overall conclusion from the comparison of 1984 and 1990 data is that the rate of hedgerow *removal* between 1984 and 1990 is greater than that in the period 1978 to 1984. In addition, there has been an overall decline in the intensity of hedgerow *management* between 1984 and 1990, leading to an increase in the boundary type defined as relict hedgerow.

Possible sources of error

While the results of this analysis provide the most up-to-date figures available on recent hedgerow changes, caution should be used in their interpretation, as follows:

(a)   The estimates of change are derived from a sample-based survey. As with any such system, there are statistical errors associated with extrapolation from a sample to national estimates, and these should be considered when drawing conclusions from change data.

(b)   Although every effort was made to standardize recording procedures in the field (including: an extensive training course; use of a field handbook; use of aerial photographs; field supervision and checks; mixing of field teams, etc.), there are likely to be some differences in the way that the data have been recorded by different observers. There is no reason to expect estimates of hedgerow recording to be biased in any particular direction and it is likely that any differences will 'balance out' over the whole data set.

It has become apparent during the analysis of *Countryside Survey 1990* data that while the definitions given in Table 11.6 are quite adequate to describe the features in most cases, there will always be occasions when the individual surveyor has to use an element of personal judgement because the feature is at the very extremes of the given definition. Examples of the range of features that might be coded as hedges are in Appendix 11.2.

When comparing the estimates made from the Institute of Terrestrial Ecology surveys with results from other studies, it is essential that definitions of categories in each survey are thoroughly understood. It is also important to know how and when each code has been applied. For instance, the Institute of Terrestrial Ecology has not included hedges that form part of a boundary between grounds associated with buildings (curtilages) and agricultural land.

The results from *Countryside Survey 1990* were eventually announced in October 1991 in a Parliamentary statement which summarized the main results from a report published at the same time (Barr *et al.*, 1991).

Government policy on hedgerow protection from July 1991

The Minister's written statement on 25 July 1991, prepared after early indications from the Institute of Terrestrial Ecology survey results had been seen, clarified the Government's hedgerow policy and were generally well received (Anon., 1991). He described two separate measures for hedgerow protection: a 'hedgerow notification' scheme and a 'hedgerow management' scheme. The management scheme would offer grants aimed at landowners and farmers to reintroduce environmentally beneficial management, including laying, coppicing and, where appropriate, biennial trimming, but not the costs of annual maintenance. This proposal accepted that in some cases the most environmentally beneficial course of action might be to leave a hedgerow unmanaged for the immediate future.

The notification scheme would require those who wished to remove hedgerows to notify their local planning authority. The local authority would then have the right to refuse permission to remove a hedgerow if it considered it to be of value for landscape, wildlife or for historical reasons. Unless there was a successful appeal to the Secretary of State against this decision, the hedge would then be registered and remain legally protected against future damage or removal.

Implementation of government policy on hedgerows

The hedgerow management scheme was launched by the Countryside Commission as the Hedgerow Incentive Scheme in July 1992 (Whelon, 1994). Regional targeting of this scheme was informed by the results from *Countryside Survey 1990* and by a more detailed analysis of the hedgerow data which provided a classification of hedgerows showing the relative distribution of particularly diverse hedgerow types (Cummins *et al.*, 1992).

The hedgerow notification scheme requires legislation before it can be put into effect and the Government plans to act on this when parliamentary time permits.

However, the lack of progress by October 1992 prompted the tabling of a 'Private Member's Bill' on hedgerow protection. The passage of this Bill was eventually unsuccessful and, as of August 1993, hedgerow protection remains on the Government's agenda for future legislation without any clear indication of when this might be.

Future research

The hedgerow story is not yet complete and research will continue to make a contribution to the development and implementation of hedgerow protection policy. This research is likely to be in four main areas:

- *Monitoring* of hedgerows to evaluate the effectiveness of current measures. These need to be conducted to rigorous standards and provide results comparable with those from *Countryside Survey 1990*.
- *Targeting* criteria and methods need to be developed which provide methods of identifying 'key' hedgerows in terms of landscape, wildlife and amenity value. These criteria must be worked into practical guidelines for use by local authorities at such a time as the hedgerow notification scheme becomes law. One possible approach is the Hedgerow Evaluation and Grading System (HEGS) proposed by Tofts and Clements (1994); the relationship between HEGS and the Institute of Terrestrial Ecology methods and classifications (Cummins *et al.*, 1992) needs to be considered.
- *The causes of hedgerow change* need to be clarified, particularly in socio-economic terms, so that the best and most cost-effective ways of encouraging landowners to take-up hedgerow management can be found.
- *Cost-effective management* regimes which maximize environmental benefits need to be determined.

There are several opportunities for further work associated with the hedgerow data collected in *Countryside Survey 1990*. These include:

- estimation of regional statistics;
- integration and cross-referencing with hedgerow data from other sources, e.g. results from the 'Monitoring landscape change' project, and work being undertaken by Dr Hooper at the Institute of Terrestrial Ecology Monks Wood;
- correlation with other types of data collected in the Institute of Terrestrial Ecology sample squares (e.g. land cover, vegetation, trees etc.) to characterize both the hedges and geographical regions in terms of species diversity, environmental quality, and nature conservation and landscape value;
- correlation with socio-economic data to determine the causes for identified changes in hedgerows;
- the use of pattern analysis to assess the biological importance of hedges in the countryside.

Not all of this work is likely to be funded by the Department of the Environment, but if the results are to continue to be relevant to the evolving policy issues it is important that the direct link between research and policy which has been such an important part of the development of hedgerow protection initiatives in recent years is maintained.

References

Anon. (1991) *Saw points: protecting hedges, trees, woodlands and orchards.* Council for the Protection of Rural England, Council for the Protection of Rural Wales and Council for National Parks. 40pp.

Barr CJ. (1990) Countryside survey 1990. *NERC News* no. **15**, 4-6.

Barr CJ, Bunce RGH, Clarke RT, Fuller RM, Furse MT, Gillespie MK, Groom GB, Hallam CJ, Hornung M, Howard DC, Ness MJ. (1993) *Countryside survey 1990.* Main report. London: DoE.

Barr CJ, Ball DF, Bunce RGH, Whittaker HA. (1985) Rural land use and landscape change. *Annual Report, Institute of Terrestrial Ecology 1984,* 133-5.

Barr CJ, Benefield CB, Bunce RGH, Ridsdale HA, Whittaker M. (1986) *Landscape changes in Britain.* Huntingdon: Institute of Terrestrial Ecology.

Barr CJ, Howard DC, Bunce RGH, Gillespie MK, Hallam CJ. (1991) *Changes in hedgerows in Britain between 1984 and 1990.* Report to DoE. Merlewood Research Station, Grange-over-Sands: Institute of Terrestrial Ecology.

Batho J. (1990) *Review of tree preservation policies and legislation.* Report to the Secretary of State for the Environment. London: Department of the Environment.

Bunce RGH. (1979) Ecological survey of Britain. *Annual Report, Institute of Terrestrial Ecology 1978,* 74-5.

Bunce RGH, Barr CJ, Whittaker HA. (1983) A stratification system for ecological sampling. In: Fuller RM, ed. *Ecological mapping from ground, air and space.* Institute of Terrestrial Ecology symposium no. **10.** Cambridge: Institute of Terrestrial Ecology, 39-46.

Cummins R, French D, Bunce RGH, Howard D, Barr CJ. (1992) *Diversity in British hedgerows.* Report to the Department of the Environment. 75pp.

Department of Environment (1989) *Digest of environmental protection and water statistics,* No. **12.**

Department of the Environment (1990) *This common inheritance - Britain's environmental strategy* - Environment White Paper. London: HMSO.

Hooper MD. (1968) The rates of hedgerow removal. In: Hooper MD, Holdgate MW, eds. *Hedges and hedgerow trees.* Monks Wood symposium no. **4.** Abbots Ripton: Monks Wood Experimental Station, 9-11.

Hooper MD. (1992) *Hedge management.* Institute of Terrestrial Ecology Contract Report to the Department of The Environment. 41pp.

Howard DC, Barr CJ. (1991) Sampling the countryside of Great Britain: GIS for the detection and prediction of rural change. In: *Applications in a changing world.*

Forest Resource Development Agreement report **153.** Ottawa: Forestry Canada, 171-6.

Hunting Surveys and Consultants Ltd. (1986) *Monitoring landscape change.* Borehamwood: Hunting Surveys and Consultants Ltd.

Royal Society for the Protection of Birds (1988) *Hedgerows: still a cause for concern.* Royal Society for the Protection of Birds Conservation Review No. **2.** Sandy: Royal Society for the Protection of Birds.

Tofts RJ, Clements DK. (1994) The development and testing of HEGS, a methodology for the evaluation and grading of hedgerows. In: *Field margins - integrating agriculture and conservation.* Boatman ND, ed. British Crop Protection Council Monograph no.**58,** 277-82.

Whelon D. (1994) The Hedgerow Incentive Scheme. In: Watt TA, Buckley GP, eds. *Hedgerow management and conservation.* Wye College Press, Wye College, University of London, 137-45.

Appendix 11.1   Example of a completed 'boundaries page' from the ITE field survey booklet

Each occurrence on the map of an alpha character represents a coded description of the boundary feature, as shown in the boxes at the foot of the page, using a pre-determined set of codes. Thus, each occurrence of 'A' on the map, describes a boundary which comprises codes: 321 'hedge >50% hawthorn', 341 '>2 m high', 352 'not stockproof' and 357 'trimmed'.

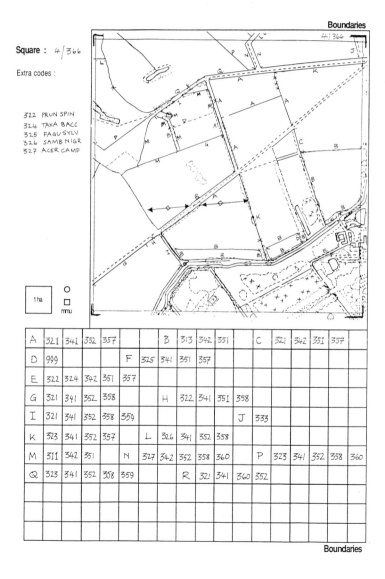

Boundaries

Square : 4/366

Extra codes :

322 PRUN SPIN
324 TAXA BACC
325 FAGU SYLV
326 SAMB NIGR
327 ACER CAMP

1ha

A	321	341	352	357		B	313	342	351		C	321	342	351	357	
D	999				F	325	341	351	357							
E	322	324	342	351	357											
G	321	341	352	358		H	322	341	351	358						
I	321	341	352	358	359					J	333					
K	323	341	352	357		L	326	341	352	358						
M	311	342	351		N	327	342	352	358	360		P	323	341	352	358	360
Q	323	341	352	358	359		R	321	341	360	352						

Boundaries

Appendix 11.2   Diagrammatic representation of different types of boundary feature that a surveyor might be required to code

A = hedge trimmed; B = hedge uncut with filled gaps <10%, not stock-proof; C = hedge derelict (or lines of shrub?); D = hedge derelict; E = hedge on bank (or line of shrubs on bank?); F = line of relict hedge (and line of trees).

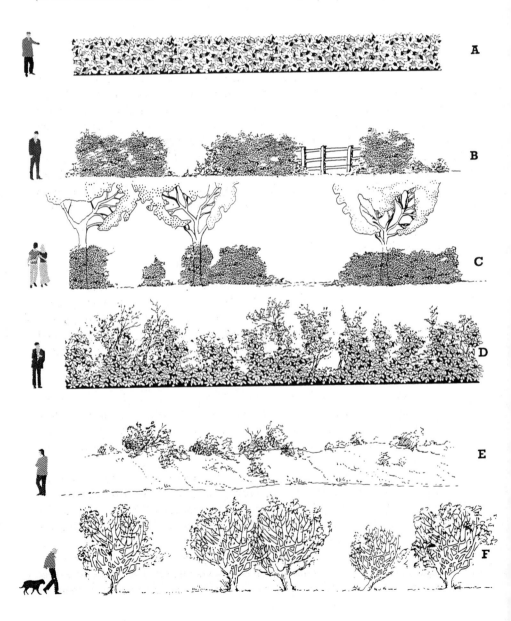

CHAPTER 12

THE HEDGEROW INCENTIVE SCHEME

D. Whelon
Countryside Commission, South West Regional Office, Bridge House, Sion Place, Clifton Down, Bristol, BS8 4AS

Introduction

The Hedgerow Incentive Scheme (HIS) was launched by the Countryside Commission in July 1992, in direct response to the trend in hedgerow decline identified by the Institute of Terrestrial Ecology (Barr and Parr, 1994). The scheme is unusual in being directly based on such empirical research. The HIS is aimed entirely at the restoration and rejuvenation of existing hedgerows and grant is not available for planting new hedges. It draws in additional funding from the Farm and Conservation Grant Scheme run by the Ministry of Agriculture Fisheries and Food where the applicant is eligible for that grant. The scheme is run from the seven regional offices of the Countryside Commission, but relies heavily on local advisers and consultants providing on-farm advice.

The policy background

Government influence on British hedgerows can be traced back to the Enclosure Acts of the seventeenth and eighteenth centuries. However, more recent government policies have, by and large, had a less positive impact on the quality and quantity of Britains hedgerow stock. The drive for increased food production during and after the Second World War led to agricultural policies which, not only gave no protection to hedgerows, but which often encouraged their removal in the pursuit of more efficient farming. Until 1970 grant was available under the Farm Improvement Scheme (run by the Ministry of Agriculture Fisheries and Food) for the removal of hedgerows although the funds spent on this were relatively small. There are no figures on the decline in hedgerow quality due to lack of appropriate management at this time, but the lack of any incentives to manage hedgerows (other than farming ones) would suggest that this may already have been a problem.

Figures released in 1986 by the Department of the Environment (Department of the Environment, 1986) showed the drastic decline in hedgerows since 1947 (from 2,600 miles per year to 1969 up to 4,000 miles per year between 1980-1985); a survey by the Ministry of Agriculture, Fisheries and Food put the figure at 1,000 miles a year lost between 1980-1985 (Ministry of Agriculture, Fisheries and Food, 1985) with half that length being replanted. Despite these figures, the physical removal of hedgerows can only be legally prevented in a few situations such as on notified Sites of Special Scientific Interest or on overgrown hedgerows covered by Tree Preservation Orders. It could be argued that peripheral legislation, such as the ban on stubble burning introduced in 1992 has had a positive effect on hedgerows by preventing damage to them.

Primary hedgerow policies have, instead, concentrated on grant aiding the planting of new hedgerows and, more recently, as the problems of lack of management have become apparent, on the restoration of existing ones. Early schemes such as the Landscape Conservation Grants (administered by the Countryside Commission and run largely through Local Authorities) and the Farm and Conservation Grant Scheme (run by the Ministry of Agriculture, Fisheries and Food) tended to look at individual landscape features, such as hedges, in isolation from the rest of the farm. More recently, however, Environmentally Sensitive Areas (run by MAFF) and Countryside Stewardship (run by the Countryside Commission) have developed a more whole farm or whole landscape approach to hedgerows. Hedgerow loss, through physical removal or lack of management, has continued despite the introduction of grant schemes.

The 1990 Environment White paper *'Our common inheritance'* gave a commitment to the further protection of British hedgerows. Original proposals for a system of Hedgerow Management Orders with accompanying compensation payments (Batho, 1990) received widespread criticism and the proposals were dropped although the Government remains committed to a system of hedgerow notification. In the meantime, attempts to create such a system of statutory protection have centred on a succession of Private Member's Bills (further details in Wilson, 1994).

Entirely separate to the above proposals, the Countryside Commission was asked by the Department of the Environment to initiate a scheme which would address the problem of the decline in existing hedgerows due to lack of or inappropriate management. This problem had been identified as part of the Land Use survey of Britain run by the Institute of Terrestrial Ecology and commissioned by the Department of the Environment. The HIS was viewed, therefore, not as an alternative to hedgerow notification nor as a compensation measure, but as a system of financial incentives for the positive restoration of hedgerows in decline.

The Countryside Commissions' initial response was that an entirely separate scheme was not necessarily the best way to proceed: grants for hedgerow planting and management were already available from a range of sources and the opportunity could be taken to try and rationalize what was already available. In particular, the Farm and Conservation Grant Scheme run by the Ministry of Agriculture, Fisheries and Food (MAFF) offered percentage grants for most hedgerow restoration work and for the planting of new hedges although the trends identified by the Institute of Terrestrial Ecology research had taken place at a time when these grants were freely available to all farmers. Another alternative put forward by the Commission was the extension of the hedgerow element of its own Countryside Stewardship scheme.

Eventually a compromise solution, worked out by the two Minsters at MAFF and the Department of the Environment, resolved that the HIS grant could be used as a 'top up' to the Farm and Conservation Grant Scheme. Those farmers eligible for the Farm and Conservation Grant Scheme would receive part payment from it and the remainder from the HIS; those not eligible for MAFF grant would simply be funded entirely from the HIS. All the administrative detail would be dealt with by the Countryside Commission and MAFF.

Having been given the broad framework within which to work, the Commission had six months to develop the scheme in consultation with the Department of the Environment and the Institute of Terrestrial Ecology and its key partners (English Heritage, English Nature and MAFF) with the views of other bodies being sought as necessary.

Whilst these consultations were taking place a market research exercise was also commissioned which sought to discover the attitudes of farmers and other hedgerow owners to their hedges, and the levels of payment which might encourage them to consider appropriate restoration management. Although a fairly small sample was selected the indications were that interest in hedgerow restoration was very high with the actual payments required to achieve results being surprisingly low. A large proportion of farmers admitted to being unaware of traditional methods of hedgerow management and placed a lot of emphasis on receiving advice. The findings of this market research were incorporated as central elements of the HIS.

The Countryside Commission were given £3.5 million to run the scheme over the first three years. With the anticipated funding from MAFF it was estimated that some £4.3 million of government money would be put into the scheme in total - enough to restore approximately 1,200 miles of hedgerow.

Developing targets

A very clear question emerged at an early stage in discussions with both the Institute of Terrestrial Ecology and the Department of the Environment: whether or not it would be possible to use the ITE data to locate the hedgerows in most need of restoration management, both in order to target scarce resources effectively and to achieve practical results on the ground. Regrettably this proved impossible for a number of reasons.

The ITE trend was based on a limited sample of approximately 1% of British hedgerows. The national trend in hedgerow decline could be clearly identified from this research, but regional or county figures were less reliable.

The Countryside Commission invariably had to make subjective decisions. For example, it was recognized that more hedgerows were in a state of decline in south-west England than in parts of East Anglia, simply because there were higher densities of hedgerows. Decisions had to be made as to the importance of hedgerows in the local, county or regional landscape and this the ITE survey could not do. Nor could it be used to decide whether the effort should be concentrated on neglected hedges or ones which had been 'over-managed'.

The ITE survey concentrated largely on species composition and diversity, both within the hedge and along its margin. From an early stage the Commission had been charged to look at four types of hedgerow as primary targets for restoration. These were:

1. Hedgerows of wildlife value either because of the presence of rare species, a particular diversity of plant life and/or because they provided links to other areas of wildlife value.
2. Hedgerows of historic interest either because of their great age (medieval field boundaries, parts of old field systems, parish boundaries etc.) or because they were a

particularly good example of an historical period (some enclosure hedges, therefore, were not ruled out).

3. Hedgerows in degraded landscapes whose restoration would also help restore and enhance those landscapes. The urban fringe and some areas of intensive arable farming were the two prime targets developed.

4. Hedgerows of amenity value particularly those in areas of high public use, alongside rights of way, canals etc. or which were highly visible from surrounding areas.

Given this wide remit information on species composition alone could not be used as a targeting mechanism. Subsequent attempts to devise a coherent national system of geographical targeting proved difficult. It became clear that, although hedgerows are a much cherished aspect of our landscape, remarkably little is known about their distribution and status.

Working to broad national guidelines it was left to the seven regional offices of the Countryside Commission, in consultation with local partner bodies, to identify appropriate targets within the four broad areas outlined above. From this a number of quite specific targets were selected such as the south-western hedgerows harbouring the rare cirl bunting and the historic hedgerows of the Ribble Valley; in other cases it was not possible to be so specific.

The Hedgerow Incentive Scheme - how it operates

Full details are contained in the *Handbook for the Hedgerow Incentive Scheme* (Countryside Commission, 1992), but the key features of the scheme are:

- The scheme is open to anyone able to enter into a ten-year agreement and applicants are encouraged to seek local advice before putting forward an application. This is to try and ensure that local variations in management are maintained and that the quality of the applications remains high.

- The application takes the form of a plan on which are marked all the hedges on the holding. Also included is a brief conservation assessment of the hedgerows including comments on their wildlife, historical and amenity value.

- Those hedgerows requiring restoration management are identified and the required management defined for a particular year. All hedgerows to be restored must have a programme of one or more of the following management methods: laying, coppicing or gapping up. A range of ancillary works, such as fencing and repair of stone-faced earth banks, can also be funded. All restored hedgerows are required to have a one-metre unsprayed, unfertilized and uncultivated strip maintained on either side of them.

- The remaining hedgerows on the holding (those not requiring restoration management) have to be managed to certain minimum standards to ensure that the problem of inappropriate management is not just rotated around the farm. These 'cross compliance' standards include controls on timing of trimming (restricted to winter months), frequency of trimming (on rotation with each hedge trimmed twice in five years) and hedge height (a minimum of two metres). Variations to these are negotiable and it is accepted that, in some cases, the minimum standards set may not be appropriate (for example, roadside hedges often need trimming annually for safety reasons).

- Applications are accepted at the discretion of the Countryside Commission. Successful applicants are offered ten year agreements, payments being made in the year that any restoration work is carried out and using standard costings for each item; a further payment is made after the fifth year of restoration, provided the work has been successful.

The Hedgerow Incentive Scheme - what makes it different

It is important to emphasize the aspects of the Hedgerow Incentive Scheme which set it apart from other hedgerow schemes. These are:

- The HIS is a discretionary scheme and applications are subject to negotiation. No work can be paid for until an agreement has been offered.
- It operates on a whole farm basis. All hedgerows on a holding are brought into some form of beneficial management.
- A plan of works is put forward which will lead to the restoration and positive management of all hedges over a full ten-year period. Agreement prescriptions are suitably flexible to allow for local variation in traditional management techniques.
- An additional payment is made after five years on restored hedges where the restoration management has been successful.
- Additional payments are made for hedges which are particularly difficult to restore either because they are very wide or they have wire entangled in them. Payments are also available for the management of hedgerow trees.
- Although the HIS is not an experimental scheme the above elements are novel and their effectiveness will be carefully monitored both during normal compliance checks and by a national Monitoring and Evaluation contract.

Operational details

The HIS is a national scheme operating throughout England and with a set of broad management prescriptions which it seeks to apply in all parts of the country; however, it is accepted that local variation in hedgerow management is important.

Each regional office of the Countryside Commission has one Hedgerow Incentive Scheme Officer whose role is to promote the scheme, to liaise with and advise local consultants and to conclude ten year HIS agreements. It is NOT, however, their primary role to advise potential applicants directly on the scheme, the emphasis being on local advisers from the private and public sector to fulfil this function; in practice these have included advisers from the Farming and Wildlife Advisory Group, the Agricultural and Development Advisory Service and a range of private consultants. Those applicants accepted into the scheme can claim for the cost of employing a consultant to help them with their application.

The link with the MAFF grant, although entailing greater administrative effort, places little additional burden on the applicant as only one application form is required.

Response to date

The HIS was launched on 20 July 1992 by the then Environment Minister, David McClean. The restoration work agreed for the first year of the scheme are summarized in Table 12.1 together with more detailed figures for the first five years (Table 12.2). A Monitoring and Evaluation contract (let to Travers Morgan) will establish the effectiveness or otherwise of the scheme after its first three years, but a number of broad comments can be made based on the evidence from Year 1.

1. There has been an extremely encouraging response to the scheme with a total of 411 ten year agreements concluded from almost 600 applications.
2. There has been variation in levels of interest both between regions and within regions. The majority of agreements have been concluded in the regions with the greatest densities of hedgerows although the south east has had a relatively small response compared to its size and number of eligible hedgerows. In the south west - the region with the highest demand - one county, Wiltshire, had only a trickle of applications. Reasons put forward by farmers and partner bodies in the areas of low demand have been:
    *   *Low payment rates:* standard levels of payment operate throughout the country but contractor rates vary considerably making the rates more attractive in some areas than others.
    *   *Size of holding:* large holdings have to commit themselves to more work than small ones over the ten years of the agreement; the management of 'cross compliance' hedgerows has been cited as a particular problem in this respect.
    *   *Arable farms:* arable farmers have little practical use for their hedgerows and, in many cases, have big farms as well. Whilst stock farmers also benefit from getting their hedges fenced, the arable farmers cite the fact that they will lose a one-metre strip of productive field if they restore their hedges under the scheme.
3. The early concerns about finding a national targeting mechanism have remained. The four broad criteria put forward initially have been useful but they are so broad that most hedges qualify under one or more of them. Attempts at geographic targeting have been tried in the regions with some success, but the problem remains that there is no national survey information which can be used to inform local targeting.
4. The principal agent of delivery - a regional Hedgerow Incentive Scheme Officer acting as the catalyst for local consultants to deliver schemes - has worked well in some regions and this would appear to be an effective way of getting local traditions of hedgerow management included in a national scheme. In some areas, however, local consultants have either not been available or have not been sufficiently attracted by the payment offered (even though that payment is made to the applicant, allowingthe consultant to charge a fully commercial rate).

Table 12.1   Hedgerow Incentive Scheme agreements 1992 – restoration work agreed for first year (October 1992 - September 1993)

	North	Midlands	South-west	Yorks and Humberside	North-west	South-east	Eastern	Total
Agreement nos	83	70	95	56	31	50	26	411
Gapping up (m)	19,736	6,978	7,339	5,169	2,494	4,766	6,850	53,332
Hedge laying (m)	22,478	9,325	19,279	4,641	5,746	8,815	1,942	72,227
Coppicing (m)	4,266	6,066	15,008	4,079	765	9,371	8,698	48,253
Totals (m)	46,480	22,369	41,626	13,889	9,005	22,952	17,490	173,812

Table 12.2 Hedgerow Incentive Scheme agreements 1992 - work   agreed for October 1992 - September 1997

	North	Midlands	South-west	Yorks and Humberside	North-west	South-east	Eastern	Total
Agreement no.	83	70	95	56	31	50	26	411
Gapping up (m)	48,058	21,897	30,477	22,120	9,807	12,019	14,546	158,925
Hedge laying (m)	62,233	55,632	76,275	25,055	22,798	27,806	8,310	278,109
Coppicing (m)	12,142	24,390	9,725	13,116	3,718	24,507	30,158	117,756
+ Wide hedges (m)	26,303	31,610	49,685	7,198	9,066	25,337	10,237	159,436
Fence removal (m)	53,443	44,113	53,077	23,550	16,605	30,404	3,607	224,800
No. hedge trees	2,153	608	357	513	45	254	445	4,375
3-line wire (m)	11,969	11,415	30,453	8,608	3,639	10,145	865	77,094
Woven wire (m)	106,673	87,796	136,912	47,004	37,963	27,747	4,769	448,864
Rabbit net (m)	7,945	3,464	7,175	3,922	3,347	3,805	406	30,064
No. pollard trees	38	47	173	3	-	123	39	423
No. tree surgery	199	132	43	118	3	170	17	682
Earth banks (m)	8,050	307	25,983	858	20	596	350	36,165
Stone banks (m)	1,074	112	2,426	130	417	-	-	4,160

5. It is too early to say to what extent the whole farm approach has acted as a barrier to potential entrants although there is some evidence that the conditions put on the management of hedges not being restored are too onerous for some.Agreement holders, however, cite the value of being able to plan ahead with a set schedule of works which is valid for ten years.

6. In some areas there have been strong demands for paying grant only for fencing, particularly where the hedge has suffered from excessive flailing. Whilst sympathetic to this problem the Countryside Commission has to be wary of those who would take advantage of such a condition to fence their fields without managing the adjacent hedges.

7. Although the scheme budget is small (£3.5 million over three years) it has gone a remarkably long way in the first year. By careful targeting of the funds available, and tying in with the other available grants wherever possible, the Countryside Commission aims to demonstrate the effectiveness of the scheme in the hope that further funding will become available in the future. It must also be stressed that, under the cross compliance element, a significant length of other hedgerow will be brought into protective management and the trend of hedgerow loss identified by ITE will at least be stemmed.

References

Batho J. (1990) *Review of tree preservation policies and legislation.* Report to the Secretary of State for the Environment. London: Department of the Environment.

Barr C, Howard D, Bunce R, Gillespie M, Hallam C. (1991) *Changes in hedgerows in Britain between 1984 and 1990.* Report to DoE. Merlewood Research Station, Grange-over-Sands: Institute of Terrestrial Ecology.

Barr CJ, Parr TW. (1994) Hedgerows: linking ecological research and countryside policy. In: Watt TA, Buckley GP, eds. *Hedgerow management and nature conservation.* Wye College Press, Wye College, University of London, 119-36.

Countryside Commission (1992) *Handbook for the Hedgerow Incentive Scheme.* (CCP 383). 8pp.

Department of the Environment (1986) *Monitoring landscape change.* Merlewood Research Station, Grange-over-Sands: Institute of Terrestrial Ecology.

Ministry of Agriculture, Fisheries and Food (1985) *Survey of environmental topics on farms.* 21pp.

Ministry of Agriculture, Fisheries and Food *(1989) Farm and conservation grant scheme handbook.* 40pp.

Wilson A. (1994). The fight for hedgerow protection legislation.In: Watt TA, Buckley GP, eds. *Hedgerow management and nature conservation.* Wye College Press, Wye College, University of London, 146-9.

Author's Note: Since writing this report two amendments need to be added:

1. After the second year of the HIS there are now over 900 ten year agreements nationally with over 1,000 miles of hedgerow due to be restored.

2. From May 1994, the HIS has been incorporated into Countryside Stewardship, also run by the Countryside Commission.

CHAPTER 13

THE FIGHT FOR HEDGEROW PROTECTION LEGISLATION

A. Wilson
Council for the Protection of Rural England, 25, Buckingham Palace Road, London, UK
Current address: *Northumberland National Park, National Park Headquarters, Eastburn, South Park, Hexham, Northumberland NE46 1BS, UK*

Introduction

Throughout the post-war period of farming expansion and intensification Government policy backed hedgerow removal. New, larger machinery needed bigger fields to operate at its most economic; farm size and scale increased; livestock farming retreated as European Community subsidies promoted arable crops; the hedgerow was seen as an encumbrance, a sign of outdated methods, a barrier to modern efficient farming.

Grants were available for hedge removal (though, curiously, grant was also available for hedge planting from the 1950s) and were eagerly snapped up. When the Council for the Protection of Rural England (CPRE) and other conservation bodies began to bemoan the loss of hedges and the transformation of large areas of lowland farmland there were two constant refrains. These are still heard today and have been repeated every time that campaigners have called for measures to give legal protection to hedges.

The two refrains

The first refrain is along the lines that 'the rate of loss has slowed greatly and major removals are now a thing of the past'. This can be found in documents from the Agricultural Development and Advisory Service back in the 1970s (Caborn, 1970). Time and again the claim has been shown to be untrue: there have been, and continue to be, substantial net losses of hedges (Department of the Environment, 1993); whether the rate is currently high or rather reduced is hardly the issue. The losses are mounting and threaten even the most valuable and irreplaceable of hedges.

The second canard is that hedge loss does not matter because most hedges date from the relatively recent enclosure period when people objected to their introduction and, therefore, removal today is just another phase of landscape change. Contrary to popular opinion, the majority of hedges pre-date the Enclosure Acts and many go back centuries earlier, even to Roman times. A more subtle and important flaw in this second argument is, however, the comparison of hedge loss today with hedge planting in the enclosure period. In that period the introduction of hedges could be seen as an environmental and social regression as common land, much of it unimproved, was lost to less diverse hedged landscapes of less public, if greater private, value. Today, by contrast, the hedge is frequently the main remaining feature of environmental interest on farms, with the greatest public value. The hedge can be all that is left of 'wildspace' on farms. It is no coincidence, let alone contradiction, that campaigners in different centuries have found common cause in attempting to deny, in the public interest, and later to protect, these wonderful creations. A lack of appreciation of the past and present context of landscape change is a real obstacle to achieving hedge protection and indeed other landscape protection measures.

From refrain to action

The first sign of Government action on the issue of hedge protection came in 1986 when former Department of the Environment Minister Angela Rumbold announced that the Government would be introducing Landscape Conservation Orders in National Parks: these would have brought protection to important hedges along with other landscape features. This was the result of long pressure to ensure that, at least in these most precious parts of the countryside, there were limits on landowners' rights to remove any stone wall, hedge, pond, etc. they desired. The 1981 Wildlife and Countryside Act had introduced powers to protect wildlife sites five years earlier. Ancient monuments, important trees and listed buildings had already been given statutory protection but other landscape features remained unprotected. The commitment to action was repeated in a blaze of publicity six months later by the Department of the Environment (DoE) Press Department.

In 1987, however, an attempt to legislate to protect hedges by the Private Member's Bill procedure was sunk on the rock of Ministry of Agriculture Fisheries and Food (MAFF) objections; the worst of hedgerow removal was now over, it was claimed. May 1987 saw a renewed commitment, this time in the Conservative Party election manifesto, to special protection for the landscape of National Parks. But the voters, despite re-electing the Government, must not have voted for better landscape protection...the promise was dropped in September 1987.

The indefatigables at CPRE kept up their work. In autumn 1990 Chris Patten's Environment White Paper *'This common inheritance'* contained a commitment to protect hedges; that was DoE News Release 509. DoE News Release 533 repeated the Government's commitment. Eight months later the Government, whipping hard, dismayed many of its own supporters by removing a hedge protection clause which had been inserted into the Planning and Compensation Bill. The anger at this action forced the Governments hand, but only a little bit. In July 1991, Junior Environment Minister Tony Baldry announced a new and powerful set of hedge protection proposals, including the Hedgerow Incentive Scheme and tough legislative proposals. The DoE News Release followed (Number 471)...but the legislation did not. The commitment to legislation was repeated in the second edition of the Environment White Paper...but legislation still did not appear. In fairness, the Government did co-operate over Peter Ainsworth's gallant attempt at another Private Member's Bill in 1992, but this was blocked by a minority of intransigent MPs with farming and landowning interests, and still no hedgerow legislation was forthcoming. The latest act in this shameful and cynical saga is that the

Government is definitely committed to hedge legislation, but they're undertaking a survey.to give up-to-date statistics!

If that survey does show a decline in outright removal of hedges, no doubt that would be the one compelling argument the Government would need to ditch their oft-repeated commitment for good, and leave every venerable and ivy-strewn grandee of a hedge to a quick afternoon's work with a chain saw. So much for history. So much for protecting our own back yard; other people's rain forests are so much easier.

Between 1947 and 1990 there was a net loss of 160,000 miles of hedge in England (Hunting Surveys and Consultants Ltd, 1986; Barr *et al.*, 1991).

Mechanisms for protection

A variety of mechanisms could be used to give legal protection to hedges: the chances are that legislation, if it materializes, will be along the lines of recent Government proposals. This effectively means that unless there is a threat to a hedge no administrative effort will be required and only then if the threatened hedge is a valuable one. In outline, this system of 'notification' would require landowners to inform local authorities if they intended to remove or damage a hedge. If the hedge was an important one the hedge would be 'registered' and it would be a criminal offence to damage it. Guidance would be given to local authorities on what constituted an 'important' hedge. This system clearly has many merits.

An alternative system would be to list all hedges of importance on a national register, rather as we do for listed buildings. The problem here is the scale of the exercise and who would undertake it. Much effort would be spent listing hedges which are secure. And 'importance' is often dependent on local factors that a national system would find hard to assess.

Another mechanism would be akin to Tree Preservation Orders, giving local authorities the ability to pre-identify and place orders on important hedges. This has its merits but again involves work which may be wasted as well-loved hedges are catalogued along with the vulnerable.

Another means of protecting hedges and other environmental features is to make their retention a condition of the receipt of other grants such as set-aside payments and Countryside Stewardship. The Ministry of Agriculture, Fisheries and Food calls this 'cross compliance'. The difficulty with this is that, in theory, it does provide some weak protection to hedges, but not necessarily the best ones and, of course, only those which are adjacent to land receiving certain forms of grant. It is haphazard. The reason the Government is prepared to accept this form of regulation and not a proper, effective, targetted system, is that it is incapable, under present arrangements, of enforcing it over most of the land in question, thus presenting little regulatory burden to MAFF or the farmer...and little environmental gain. Not just haphazard, but, at present, a sham.

Full marks to the DoE civil servant who dreamed up the notification system! Now that would *work*.

What next?

The Government is still committed to legislation. The problem is getting action which involves regulation at a time when regulation is supposedly being pruned (tell that to a farmer with an IACS form). There is now a real danger that hedge legislation will fall victim to 'deregulation'. The only solution is public pressure. Research, incentives and further surveys are all welcome, the incentives especially so. But if we want to be sure that any particular hedge is safe, we need a law which will protect it. Now.

References

Caborn J. (1970) *Hedges*. MAFF/Joint Shelter Research Committee.
Department of the Environment (1993) *Countryside survey 1990*.
Hunting Surveys and Consultants Ltd. (1986) *Monitoring landscape change*. Borehamwood: Hunting Surveys and Consultants Ltd.
Barr CJ, Howard DC, Bunce RGH, Gillespie MK, Hallam CJ. (1991) *Changes in hedgerows in Britain 1984 and 1990*. Contract report to DoE.

CHAPTER 14

ABSTRACTS

Establishment and management of thorn hedges in the Cambrian Mountains ESA

J. Wildig
ADAS, Pwllpeiran, Cwmystwyth, Aberystwyth, Dyfed SY23 4AB, Wales, UK

The Cambrian Mountains ESA lies in the centre of Wales and the farming within it is dominated by hill sheep. Thorn hedges first appeared in the area with the enclosure of the middle hill and fields on what are today's mountain sheep units, at the end of the eighteenth and beginning of the nineteenth century. In recent times, sheep numbers have increased dramatically and this, coupled with the fragmentation of many units through extensive afforestation on the lower slopes from the depression of the 1930s to the present time, has led to many hedges becoming relict with lines of individual hawthorn now indicating where hedges once were. Pwllpeiran is typical, consisting of the remains of several farms after extensive afforestation with relatively small areas of middle hill and fields left. However, these still show evidence of the more abundant hedges of the past.

A programme of hedge planting started in the mid-1980s on these lower areas with the objectives of providing stockproof fencing and shelter and improving their wildlife and conservation value. Early experiences soon showed that there was a range of both practical and R and D problems to be solved before there was any hope of success on what is basically an intensively stocked unit. For example, the hedge needs to be protected from the grazing animal. Two post and wire fences are needed for this and the distance between the two is critical:- too close and the quicks are nibbled and stunted; too wide and the vegetation enclosed by the two fences becomes luxurious and swamps the quicks and contributes towards very slow growth in the first years.

The MAFF (LUCRE) Group is currently funding three experiments at Pwllpeiran. Two of these are concerned with vegetation control in the area around the newly planted and the 2-3-year-old establishing hedges and the third with the management of a 5-6-year-old hedge which has been subjected to various management treatments to look at alternatives to layering in the management of such hedges for stockproof fencing and shelter.

Habitat preferences of birds in hedges and field margins

R.E. Green[1] and E.J. Sears[2]
[1] *Royal Society for the Protection of Birds, 17 Regent Terrace, Edinburgh EH7 5BN, Scotland, UK*
[2] *Royal Society for the Protection of Birds, The Lodge, Sandy, Bedfordshire SG19 2DL, UK*

Surveys of breeding songbirds and measurements of characteristics of hedgerows, field margins and adjacent land-use were carried out on about 300 kilometres of hedgerow on 46 farms in lowland England. Bird surveys were carried out in the early morning on two occasions, in spring and early summer, so that resident species and summer visitors were covered. Hedgerows were divided arbitrarily into 50 metre sections and the presence or absence of each bird species was scored as a binary variable for each section, regardless of the number of individuals seen or whether birds were seen on one or both visits.

Habitat information was recorded on a separate visit. For each hedgerow section the height and width of the hedge, the proportion of the length of the section with woody vegetation, the dominant species of hedgerow shrub, the number of shrub species in a standard length, the number of trees, the dominant species of ground vegetation, both underneath the hedge and on the uncultivated margin outwith the hedge and the area of this uncultivated margin within the section were measured. The land-use on both sides of the hedge was recorded (type of crop or grass, road margin). The farms in the study were selected to include some where there was reduced spraying of herbicide and insecticide on the margins of cereal crops ('conservation headlands').

The presence/absence of each bird species was modelled using multivariate logistic regression analysis. Categorical variables, such as dominant shrub species, were modelled as factors. Plots of bird incidence against tree number, hedge height etc. suggested bell-shaped relationships with optimum values for some bird species-habitat variable combinations, so the model selection procedure included assessing the improvement in fit resulting from using a quadratic term to describe such a relationship. Models were fitted by a step-down procedure in which all of the retained variables had a statistically significant ($P<0.05$) effect.

After excluding sections with vegetation species too rare to warrant the inclusion of extra variables in the model there were 4760 50 metre sections eligible for analysis. Models were fitted for the 18 songbird species which occurred in 30 or more sections. In all species at least one variable had a significant effect on bird incidence. Tree number had a significant effect in 11 species, hedge height in 9 species and hedge width in 7 species. In several of these relationships there was evidence of a bell-shaped curve and an optimum value. There was evidence of a positive effect of hedgerow shrub species richness for 11 species which was present when the dominant shrub species was also included in the model. Dominant shrub species influenced the incidence of 4 bird species and there was some consistency among bird species in the preference ranking of shrub species; bramble tending to be preferred and field maple tending to be avoided.

Significant effects of adjacent land-use were found for 10 bird species, but the pattern of effects varied considerably among species. A separate analysis of the effects of crop type for those hedgerows with arable crops on both sides revealed only a few significant effects. No obvious effects on bird incidence of restricted spraying of herbicides and insecticides on the margins of cereal fields ('conservation headlands') were apparent, but it should be noted that this practice was not long established on some of the farms in the study and the effect might develop over a longer period.

Field boundary management and the wildlife value of hawthorn hedges

A.C. Bell, J.H. McAdam, T. Henry
Department of Agriculture for Northern Ireland, Agricultural Botany Research Division, Newforge Lane, Belfast BT9 5PX, Northern Ireland, UK

The typical agricultural scenario in Northern Ireland is a grass-based enterprise, with field boundaries commonly delimited by hawthorn hedges. In 1990 a field experiment was established to examine the impact of different management regimes on the wildlife value of the field margin, incorporating the hedge. Three well-maintained hawthorn hedges which had separated paired grass fields were selected. The treatment plots extended along 30 metres of hedge and 10 metres into each adjacent field. Within each of three blocks there were four treatments: (1) unmanaged; (2) fertilized and rotationally grazed with sheep; (3) a 2 metre wide strip ploughed along the hedge-base and sown with a game cover crop, the remaining 8 metres harvested for silage and (4) as for the previous treatment except that the ploughed strip was left unmanaged to permit colonization by natural flora.

Flora and fauna species diversity is being monitored to determine which management option maximizes the wildlife value of the field margin. Groups of three permanent 1 square metre quadrats have been located at four distances from the hedge base. In 1991 the highest species diversity in all treatments was found between 0.5 metres and 2 metres from the hedge base, with the greatest species diversity occurring within the ploughed strips.

Pitfall traps have been placed within 1-2 metres and 8-10 metres from the hedge base, to sample species associated with the field edge, and those which are active further out into the field. Initial data indicated that species diversity of carabid beetles was higher in the hedge margin compared to grassland early in the year and that the highest diversity was found in the ploughed strips. Species counts in all treatments increased as the season progressed, although the fertilized and grazed plots lagged behind the others in this respect.

Measurements of flora and fauna over 10 years will indicate which management option is best in terms of wildlife value of grass field margins.

The composition, structure and management of hedges in Northern Ireland

C.A. Hegarty, A. Cooper

Department of Environmental Studies, University of Ulster, Coleraine   BT52 1SA, Northern Ireland, UK

Hedges are a major source of wildlife and landscape diversity in the Irish countryside. Irish hedges are generally much younger than hedges in England, the majority being planted between 1750 and 1850. Although in Northern Ireland (NI) widespread removal has not occurred on the same scale as in England, removal rates of 0.5% per year since the 1960s are common (Murry *et al.*, 1992).

Field surveys of the botanical composition, structure and management of hedges has been undertaken, sponsored by the Environment Service of the Department of the Environment (NI). The NI multivariate land classification (Cooper, 1986) provided a stratification for a random sampling program of hedges, based on 25 hectare ordnance Survey grid squares. Multivariate classifications (Hill, 1979) of four habitats (trees and shrubs; hedge bank ground flora; ditch flora; and field margin flora) were carried out. Hedges were also classified by their structural attributes.

Association statistics showed significant relationships between, for example, species-rich tree and shrub groups and species-rich ground flora groups. A woodland flora characterized the species-rich hedges, whilst competitive grasses dominated the species-poor hedges. Regional differences in the composition, structure and management of hedges were evident, with the relatively unmanaged hedges of County Fermanagh associated with the greatest species diversity. Fermanagh also has more parcels of semi-natural vegetation and is farmed less intensively than the other study areas such as lowland County Antrim and County Londonderry. This suggests that land use is a key factor influencing hedge composition. Hedge management was chiefly confined to intensive lowland farms and roads.

Canonical correspondence analysis (ter Braak, 1988) was used to explore the relationships between soil nutrients (nitrogen (N), potassium (K), phosphorus (P)), soil pH, and the ground flora groups. Soils of the species-poor group had significantly higher NPK concentrations than the species-rich groups.

Townland hedges are one of the oldest hedge types in Ireland. They were found to have a greater tree and shrub species diversity and were associated more with woodland herbs. Their greater habitat diversity may account for much of this. There was, however, no significant difference in the proportions of the hedge groups associated with townland boundaries compared to other boundary types, suggesting that land use and management are also key factors influencing the species composition of townland boundaries.

Cooper A. (1986) *The Northern Ireland Land Classification*. Report to the Countryside and Wildlife Branch. Department of the Environment, NI. University of Ulster.

Hill MO. (1970) *TWINSPAN - A FORTRAN program for arranging multivariate data in an ordered two-way table by classification of individuals and attributes*. New York; Cornell University. 90pp.

Murray R, Cooper A, McCann T. (1992) Monitoring environmental change: a land classification approach. In: Cooper A, Wilson P, eds. *Managing land use change*. Geographical Society of Ireland Special Publication. No. 7. Coleraine.

ter Braak CJF. (1988) *CANOCO-A FORTRAN program for canonical correspondence analysis, principal components analysis and redundancy analysis*. Wageningen, The Netherlands: Agricultural - Mathematics Groups.

New farm hedges for wildlife and coppice

J.M. Walsingham, P.M. Harris
Department of Agriculture, University of Reading, Earley Gate, PO Box 236, Reading RG6 2AT, UK

The creation and re-establishment of hedges on farms is important for the provision of habitats for wildlife; it can usually only be achieved with the goodwill of farmers and with the minimum of expense.

A new hedge has been established on the University of Reading's Sonning Farm which is designed to be able to be coppiced and thus provide some return to the farm. Ash and poplar were used, but the ash showed such poor growth on the site that it was replaced by birch and red alder. The area between the double-row hedge and either side of the hedge was planted with wildflower seed mixes. The arable crops on either side of the hedge are managed with either full agricultural inputs or with no inputs at all.

The aims are to monitor the value of the hedge as a wildlife habitat, the effect of the hedge on the neighbouring crops and the effect of the crop management on the hedge.

Establishment and management of herb-rich field margins

E. J. P. Marshall[1], M. Nowakowski[2]
[1] *Department of Agricultural Sciences, University of Bristol, AFRC Institute of Arable Crops Research, Long Ashton Research Station, Bristol BS18 9AF, UK*
[2] *Willmot Conservation, West Yoke, Ash, Wrotham, Kent TN15 7HU, UK*

Field boundaries are an integral part of farm landscapes in lowland Britain. They form an important refuge for many species, some of which interact with adjacent arable crops. The proposed changes in price support for arable crop production, together with compulsory 'set-aside', may offer greater opportunities than in the past for the creation, or re-creation, of herb-rich strips around field edges. Diverse field margin strips may satisfy conservation and agricultural objectives in farmland by creating habitat, reducing weed ingress by out-competing annual species, and encouraging populations of predators of crop pests.

In order to examine species establishment and the requirements for subsequent management of diverse sown strips, a mixture of grasses and wild flowers was sown along the edge of an arable field in September. A range of cutting and herbicide treatments was applied to randomized plots during the following season. Main plot treatments comprised (1) unsown set-aside and, on sown plots, (2) an untreated control, (3) a single cut, (4) repeated mowing, (5) benazolin + clopyralid, (6) fluazifop-P-butyl + quinmerac, (7) cycloxydim, (8) alloxydim-sodium and (9) mefluidide. In the second year, after all plots had been mown in September, plots were split, half being mown again in March and half treated with fluazifop-P-butyl for the control of grass weeds.

Table 14.1   Numbers of sown and unsown species per main plot in the second season after sowing, on field margin plots receiving different treatments

Main plot treatment	Sown species			Unsown species			
	Grasses	Dicotyledonous		Grasses	Dicotyledonous		
		Perennials	Annuals		Perennials	Annuals	Biennials
1	1.50	2.67	1.67	5.67	2.00	1.50	5.33
2	6.67	11.00	1.83	1.00	0.00	0.50	0.67
3	6.00	11.50	0.83	2.00	0.83	0.67	1.33
4	5.33	12.00	1.17	0.83	0.33	0.33	1.17
5	5.67	8.00	1.50	1.17	0.17	0.17	1.17
6	6.33	12.67	1.17	1.00	0.33	0.50	0.83
7	5.50	13.50	2.00	0.83	0.17	0.33	2.00
8	5.50	13.83	2.17	1.50	0.17	0.67	1.33
9	5.67	10.33	1.67	2.33	0.33	0.33	1.67
SED df=15	0.762	1.120	N.S.	0.959	0.311	0.330	0.430
Overall mean	5.35	10.61	1.56	1.81	0.48	0.56	1.72

In the first year, regular mowing reduced diversity of sown annual species, while benazolin + clopyralid reduced sown dicotyledonous species. In the second year, considering the main plot effects, there were increased numbers of sown species in undrilled plots, spreading from adjacent areas. In addition, unsown perennial and biennial dicotyledonous species increased on unsown set-aside plots. On the drilled plots, there were no significant differences in the numbers of sown or unsown grasses (Table 14.1). Benazolin + clopyralid-treated plots supported reduced numbers of sown dicotyledonous species. Numbers of sown and unsown annual dicotyledonous species were generally lower in 1991, compared with 1990. Numbers of unsown grass species also declined on drilled plots. In comparison to the March cut, fluazifop-P-butyl enhanced the diversity of sown dicotyledonous species, while reducing the mean number of grass species found per quadrat (Table 2).

Table 14.2   Mean numbers of species recorded per quadrat on field margin strip sub-plots receiving a cut in March or an application of fluazifop-P-butyl

Sub-plot	Sown species			Unsown species			
	Grasses	Dicotyledonous		Grasses	Dicotyledonous		
		Perennials	Annuals		Perennials	Annuals	Biennials
Cut	4.25	3.70	0.35	0.62	0.12	0.21	0.70
fluazif.	3.72	4.22	0.38	0.39	0.10	0.15	0.83
SED df=18	0.094	0.159	N.S.	0.070	N.S.	0.027	N.S.

In conclusion, seed mixtures were established successfully with a range of cutting and herbicide treatments. Graminicides were most successful in encouraging sown dicotyledonous species.

Marshall EJP, Nowakowski M. (1991) The use of herbicides in the creation of a herb-rich field margin. *Proceedings 1991 Brighton Crop Protection Conference - Weeds,* 655-60.
Marshall EJP, Nowakowski M. (1992) Herbicide and cutting treatments for establishment and management of diverse field margin strips. *Aspects of Applied Biology* **29,** *Vegetation Management in Forestry, Amenity and Conservation Areas,* 425-30.

Hedgerow studies at Kingcombe Meadows Nature Reserve, Dorset

D. Elton
c/o Dorset Trust for Nature Conservation, Half Moon House, 15 North Square, Dorchester DT1 1HY, UK

A complex network of hedgerows, many overgrown and now largely composed of tall trees, is a striking and important landscape and ecological feature of the 600 acre Lower Kingcombe estate, 373 acres of which is now owned and managed by the Dorset Trust for Nature Conservation. Comparison with the tithe map of *c.*1840 (the earliest known map of the parish) shows that the hedge network has remained largely intact since that time. However, some hedges are clearly very old, containing large coppice stools and laid trunks, and may be medieval or earlier in origin.

The present study aims to record the structure, form and species composition of this important historical feature and wildlife habitat. The information gathered will help identify the most significant hedges — botanically, structurally, historically or in other ways — and will assist in deciding future management policy and priorities.

Each field boundary is surveyed using a pro-forma recording sheet in conjunction with a large-scale A4 map of the relevant field. Significant physical attributes, tree/shrub composition and species number per 30 yard length, canopy trees, growth pattern, banks, ditches etc., are recorded, and relevant features are mapped. The recording sheet has been designed for use by non-expert volunteers and requires the identification of trees, shrubs and climbers, but not ground flora. (This is being recorded separately by Trust volunteers on a field-by-field basis.) Hedgerows of especial interest for lichens and bryophytes are noted and these will be recorded by specialists.

The results will be summarized in map and tabular form and subject to appropriate analysis. Of particular interest will be the distribution and location of 'double' hedges which are a particular feature of the reserve; those with significant numbers of hedgerow trees; the variation in species composition between hedges and across the reserve; and the species richness which may give an indication of the possible age of the hedgerow. Other features such as growth form and overall size are likely to be of particular relevance to the value of the hedge for nesting birds and are the subject of a separate study. Hooper's method will be used to attempt to date the hedges, comparing the results where possible with a previous study, and relating them to current archaeological and landscape history studies which provide other evidence of age and origin.

Artificial overwintering habitats for polyphagous predators

A. MacLeod
Biology Department, Southampton University, Southampton SO9 3TU, UK

Previous work at Southampton has shown that within two years of creating an artificial habitat within a cereal field in the form of a raised grassy bank, simulating a grassy hedge bottom, overwintering densities of beneficial predators up to ten times greater than those which occur in naturally existing hedgerow bottoms can be obtained. It has been shown that the carabids and staphylinids which overwinter in the banks move out into the field in the spring, providing penetration of the crop early in the season during a critical period in the development of aphid populations. The overwintering beneficial arthropod community composition of the banks has been assessed previously between 1987-1990.

This work continues to monitor the overwintering densities of carabids, staphylinids and spiders in two experimental ridges. The ridges consist of four grass species sown in single species treatments and as mixtures. Predator densities were measured using samples taken by digging turves from the experimental single species plots. Two turves per plot were dug.

Staphylinids show a trend over each of the treatments although Lycosidae have increased in proportion to Linyphiidae in each grass. Total carabid density has declined in all grasses over the period of study. Trends in the fluctuations of arthropod numbers will be discussed.

Use of *Phacelia tanacetifolia* borders to enhance hoverfly populations in winter wheat

J. Hickman
Biology Department, Southampton University, Southampton SO9 3TU, UK

Adult hoverflies feed on pollen and nectar. In many species the larvae are aphidophagous with the potential for biocontrol. As a result of the removal and mis-management of hedgerows and of herbicide drift, many field boundaries are impoverished in terms of the flowering plants that they support. As a result, food resources for adult hoverflies may be limiting. The main aim of the project described below was to find whether the provision of boundary strips of the North American annual *Phacelia tanacetifolia* (Benth.) (Hydrophyllaceae) as a pollen resource for adult hoverflies resulted in increased biocontrol of aphids in the adjoining fields compared with control fields. To assess this the number of adult hoverflies, eggs and aphids in *Phacelia*-bordered and control fields were compared.

A 0.5 metre strip of *Phacelia* was hand drilled in mid-March 1992 between the crop and the field boundary along two adjacent sides of each of three fields of winter wheat on a farm in North Hampshire. Three other fields without *Phacelia* were used as controls.

Fluorescent yellow water traps (painted 19 centimetre plastic plant pot holders) were put out at crop height at different distances from the border in experimental and control fields; trapped hoverflies were removed from these traps at regular intervals from the time that the *Phacelia* started to flower at the beginning of June until it became senescent in August. Pots of barley seedlings infested with aphids (*Sitobion avenae* (Fabricius)) were put at the base of each trap; these were changed every two days and the number of hoverfly eggs on the barley counted. The percentage of wheat stems with one or more aphids was calculated once a week in the vicinity of each trap.

Significantly more hoverflies were trapped in the *Phacelia*-bordered fields than in the control fields and significantly more oviposition was detected. No significant differences were found in aphid numbers in the crop between experimental and control fields. This may have been due to the fact that by the time that large numbers of hoverflies were entering the crop, the wheat, which matured approximately two weeks earlier than usual, had passed the most suitable stage for aphids and numbers were declining in all fields.

These results show that the provision of flowering strips of *Phacelia* does enhance hoverfly populations and oviposition rate in the crop; the effect of this on aphid numbers in wheat remains to be demonstrated, although there is evidence from other crops that there is an effect on this parameter also.

Management of hedgerow vegetation for weed control and enhancement of beneficial insects

F. Rothery

Biology Department, Southampton University, Southampton SO9 3TU, UK

Many hedgerows have become a problem source of annual weeds, such as cleavers and barren brome; these have been able to establish following damage to the perennial flora of the hedge-bottom. Probable causes of this are herbicide and fertilizer drift and cultivation too close to the hedge. The creation of a dense sward of perennial grasses and herbs will reduce penetration by annual weeds, and those, such as cleavers, that can still germinate in shady bare ground at the base of the hedge, will be unable to cross the vegetation barrier and invade the field.

Three hedgerow experiments are currently being set up to evaluate different management strategies to bring about re-establishment of perennial hedgerow species, and to assess the relative effectiveness of different grasses and flowers as a resource for beneficial insects.

Experiment 1 assesses management techniques for weed control and the provision of overwintering habitats for predatory arthropods in a poor quality hedgerow. There are nine treatments, using different management techniques of cutting, selective herbicide treatment and autumn-sowing with single and mixed stands of perennial grasses. Each is replicated three times as 2 metre x 10 metre plots in a randomized block design along the length of a hedge. The botanical composition of the replicates will be assessed annually in the late spring/early summer, and overwintering predators will be sampled annually during January and February.

Experiment 2 compares some of the sown grasses used in experiment 1 with the vegetation of an undamaged hedgerow. There will be four treatments, each replicated three times in a randomized block design.

Experiment 3 compares 17 different dicotyledonous species as a resource for beneficial insects. Pure stands of each species with a grass mixture, and an additional treatment consisting of a mixture of species with a grass mixture, will be sown in the autumn of 1992. The 18 treatments will be replicated three times in 2 metre x 5 metre plots, in a randomized block design.

Comparison of the treatments will be made by carrying out census walks along the experiment, and recording flower use by hoverflies, bumblebees and butterflies. Insects will also be collected in water traps in each replicate, and numbers of beneficial insects caught will be compared, e.g. parasitoids, ladybirds and hoverflies. Hoverflies will later be dissected and their gut contents examined microscopically to assess the proportions of different pollen types taken.

The overall aim of the project is to arrive at a recommendation for restoring the perennial flora of a damaged hedgerow, in a way that improves the resources for predatory arthropods at the same time as improving the conservation value of the hedge.

The factors affecting butterfly distribution on arable farmland

J.W. Dover
The Game Conservancy, Fordingbridge, Hampshire, SP6 1EF, UK

The effects of biotic and abiotic factors on the distribution and microdistribution of three species of satyrid butterfly, ringlet, (*Aphantopus hyperantus*), the meadow brown (*Maniola jurtina*) and the gatekeeper, (*Pyronia tithonus*) were studied in a block of arable farmland on a north Hampshire farm.

Butterflies of the three species were studied in the field margins surrounding a small copse in 1988 and 1989. The field margins were divided up into 10 metre long numbered plots using flagged canes and stakes, 592 plots were assessed in this study in 1988 and 480 plots in 1989. Butterflies were captured throughout their flight period, identified to species and sex and released - after marking on the ventral wing surfaces. Biotic and abiotic habitat parameters were recorded in July for each 10 metre plot. Distribution maps and movement plots were produced from the mark-release-recapture data. A subset of the data was analysed using stepwise multiple regression to investigate the factors influencing the observed distribution patterns. Analyses were performed on whole margin data and on 30 metre long sections of field margins (data for 3 x 10 metre plots combined/section).

The principal factors affecting satyrid distribution at the level of the field margin were the presence of a small uncropped area, the presence of shelter and the abundance of nectar plants. Non-floral factors affecting microdistribution included the degree of shelter, sunlight, width of hedgebank or grass verge and a small area of uncultivated ground between the edge of a copse and an adjacent hedge. The presence of farm tracks at the field margin was shown to exert a negative influence on abundance. Floral variables including the abundance of bramble (*Rubus fruticosus*), marjoram (*Origanum vulgare*), and thistle-like Compositae were all positive factors affecting satyrid micro-distribution. Negative floral factors, possibly indicative of habitat degradation, included cow parsley (*Anthriscus sylvestris*), and old man's beard (*Clematis vitalba*). The latter was a negative factor for *A. hyperantus* and *M. jurtina*, but a positive factor for *P. tithonus*.

The influence of landscape structure, in particular the presence of hedgerows, and floral resources was discussed in relation to satyrid distribution and conservation.

Factors affecting the development of conservation headlands in grassland

D.G. Gwynne, T.H.W. Bromilow
Environmental Sciences Department, SAC, Auchincruive, Ayr, Scotland KA6 5HW, UK

Information is required on the management of conservation headlands in grassland to optimize their wildlife conservation value, in parallel with similar studies on arable land and this will provide a basis for conservation advice to farmers. The factors affecting the development of floral and faunal diversity in grassland headlands are being studied at two locations in the west of Scotland after exclusion of grazing. Treatments include nil fertilizer, cultivation, seeding with windflower mixtures, reducing soil pH and cutting. Seedbank analysis, soil parameters and invertebrate species have been recorded.

Cultivations, seeding and reduced fertilizer use have clearly had the most noticeable effects on the vegetation. The cultivated plots were initially dominated by ruderal and arable species and initial gradient of species diversity from the hedge outwards has become less evident with time.

162

INDEX

Fauna have been indexed under their English names, flora (where botanical names are used in any part of the text) in Latin, with English explanations appended. Some trees are listed only in English. Composite English names are listed uninverted, so 'bank vole' is to be found under B rather than V. 'Butterflies' and 'invertebrates' are consolidated. All references to the civil service and legislative process are listed under 'government'.

(The index was compiled by Christine Headley.)